Differential Geometry, Functional Analysis and Applications

Differential Geometry, Functional Analysis and Applications

Editors
**Mohammad Hasan Shahid • Sharfuddin Ahmad
Khalil Ahmad • M. Rais Khan
Khursheed Haider**

Associate Editors
**Shahzad Hasan • Arshad Khan
M. Yahya Abbasi**

Narosa Publishing House
New Delhi   Chennai   Mumbai   Kolkata

Differential Geometry, Functional Analysis and Applications
166 pgs.

Editors
Mohammad Hasan Shahid
Sharfuddin Ahmad
Khalil Ahmad
M. Rais Khan
Khursheed Haider

Associate Editors
Shahzad Hasan
Arshad Khan
M. Yahya Abbasi

Department of Mathematics
Faculty of Natural Sciences
Jamia Millia Islamia (Central University)
New Delhi

Copyright © 2015, Narosa Publishing House Pvt. Ltd.

NAROSA PUBLISHING HOUSE PVT. LTD.

22 Delhi Medical Association Road, Daryaganj, New Delhi 110 002
35-36 Greams Road, Thousand Lights, Chennai 600 006
306 Shiv Centre, Sector 17, Vashi, Navi Mumbai 400 703
2F-2G Shivam Chambers, 53 Syed Amir Ali Avenue, Kolkata 700 019

www.narosa.com

All rights reserved. No part of this publication may be reproduced, stored in a retrieval system, or transmitted in any form or by any means, electronic, mechanical, photocopying, recording or otherwise, without prior written permission of the publisher.

All export rights for this book vest exclusively with Narosa Publishing House Pvt. Ltd. Unauthorised export is a violation of terms of sale and is subject to legal action.

Printed from the camera-ready copy provided by the Editors.

ISBN 978-81-8487-421-1

Published by N.K. Mehra for Narosa Publishing House Pvt. Ltd.,
22 Delhi Medical Association Road, Daryaganj, New Delhi 110 002

Printed in India

PREFACE

Differential Geometry and Functional Analysis are important branches of Mathematics having many applications in different areas of Mathematical sciences. An international conference on "Differential Geometry, Functional Analysis and Applications" was held at Department of Mathematics, Jamia Millia Islamia, New Delhi under DRS-I of University Grant Commission, Govt. of India. The conference focused on some selected topics namely Submanifolds Theory, Fibre bundle, Harmonic morphisms, Homogeneous and symmetric spaces, Structures on manifolds, Variational analysis, Fixed point theory, Operator theory, Fourier analysis, Wavelet analysis, Approximation theory.

The conference was attended by participants from different parts of India and abroad.

We have received many papers for the proceeding of the conference and after due process of refereeing, some papers have been accepted for publication in this volume.

Department of Mathematics, JMI expresses its sincere thanks to the UGC, NBHM, DST, CSIR, Bank of India and officials of JMI for providing financial support to hold this conference. We are thankful to INSA for its financial support which is being utilized for the publication of this volume. Finally, we appreciate Narosa Publishing House, New Delhi for their good job in publishing this volume.

Editors
Mohammad Hasan Shahid
Sharfuddin Ahmad
Khalil Ahmad
M. Rais Khan
Khursheed Haider

LIST OF CONTRIBUTORS

ZAFAR AHSAN, Department of Mathematics
Aligarh Muslim University, Aligarh-202 002, India
E-mail: zafar.ahsan@rediffmail.com

MUSAVVIR ALI, Department of Mathematics
Aligarh Muslim University, Aligarh-202 002, India
E-mail: musavvir.alig@gmail.com

RAM SHANKAR GUPTA, University School of Basic and Applied Sciences
Guru Gobind Singh Indraprastha University
Dwarka Sec-16C, New Delhi-110 075, India
E-mail: ramshankar.gupta@gmail.com

MANISH GOGNA, Baba Banda Singh Bahadur Engineering College
Fathehgarh Sahib, India
E-mail: manish bbsbec@yahoo.co.in

SANGEET KUMAR, Department of Applied Sciences
Chitkara University, Jhansla, Patiala, India
E-mail: sp7maths@gmail.com

RAKESH KUMAR, University College of Engineering
Punjabi University, Patiala, India
E-mail: dr rk37c@yahoo.co.in

R. K. NAGAICH, Department of Mathematics
Punjabi University, Patiala, India.
E-mail: rakeshnagaich@yahoo.com

SHYAMAL KUMAR HUI, Nikhil Banga Sikshan Mahavidyalaya
Bishnupur, Bankura-722 122, West Bengal, India
E-mail: shyamal_hui@yahoo.co.in

RASHMI, School of Mathematics and Computer Applications
Thapar University, Patiala, India
E-mail: rashmi.sachdeva86@gmail.com

S. S. BHATIA, School of Mathematics and Computer Applications
Thapar University, Patiala, India
E-mail: ssbhatia@thapar.edu

MAJID ALI CHOUDHARY, Department of Mathematics
Jamia Millia Islamia, New Delhi, India
E-mail: majidalichoudhary@yahoo.co.in

MOHAMMED JAMALI, Department of Applied Sciences and Humanities
Al-Falah University, Dhauj, Faridabad, Haryana, India
E-mail: jamali_dbd@yahoo.co.in

RAJENDRA PRASAD, Department of Mathematics and Astronomy
University of Lucknow, Lucknow-226 007, India
E-mail: rp.manpur@rediffmail.com

KWANG-SOON PARK, Department of Mathematical Sciences
Seol National University, Seol-151 747, South Korea
E-mail: parkksn@gmail.com

JAI PRAKASH, Department of Mathematics and Astronomy
University of Lucknow, Lucknow-226 007, India
E-mail: drjp.0000@gmail.com

W. M. KOZLOWSKI, School of Mathematics and Statistics
University of New South Wales, Sydney, NSW 2052, Australia
E-mail: w.m.kozlowski@unsw.edu.au

AMIT SINGH, Department of Mathematics
Government Degree College Billawar, Jammu and Kashmir-184 204, India
E-mail: singhamit841@gmail.com

B. FISHER, Department of Mathematics
University of Leicester, Leicester, LE1 7RH, England
E-mail: fbr@le.ac.uk

CONTENTS

Preface — v
List of Contributors — vii

Symmetries of Type N Pure Radiation Fields and Ricci Solitons — 1
  Zafar Ahsan and Musavvir Ali

Semi-Symmetric Lightlike Hypersurfaces of an Indefinite Kenmotsu Space Form — 11
  Ram Shankar Gupta

Generic Lightlike Submanifolds of Indefinite Kaehler Manifolds — 22
  Manish Gogna, Sangeet Kumar, Rakesh Kumar and R. K. Nagaich

On the W_2-Curvature Tensor of $N(k)$-Contact Metric Manifolds — 34
  Shyamal Kumar Hui

Slant Lightlike Submanifolds of Indefinite almost Contact Manifolds — 46
  Rashmi, Rakesh Kumar and S. S. Bhatia

On Doubly Twisted Product CR-Submanifolds — 57
  Majid Ali Choudhary and Mohammed Jamali

A Note on Trans-Sasakan Manifolds — 64
  Rajendra Prasad, Kwang-Soon Park and Jai Prakash

On the Cauchy Problem for the Nonlinear Differential Equations with Values in Modular Function Spaces — 75
  W. M. Kozlowski

Coincidences and Common Fixed Point Theorems in Intuitionistic Fuzzy Metric Spaces using General Contractive Condition of Integral Type — 106
  Amit Singh and B. Fisher

Characterization of Sobolev Spaces using *M*-Band Framelet Packets   119
F. A. Shah, Humaira Siddiqui and K. Ahmad

Approximation of Bound for the Class of Polynomials Vanishing Inside the Disk   133
Arty Ahuja and K.K. Dewan

Strong and Δ-Convergence of Khan et al. Iterative Procedure in CAT(0) Spaces   143
Madhu Aggarwal and Renu Chugh

Differential Geometry, Functional Analysis and Applications
Editors: Mohammad Hasan Shahid, Sharfuddin Ahmad *et al.*
Copyright © 2015, Narosa Publishing House, New Delhi

SYMMETRIES OF TYPE N PURE RADIATION FIELDS AND RICCI SOLITONS

ZAFAR AHSAN[1] AND MUSAVVIR ALI[2]

ABSTRACT. The present paper puts an emphasis on the description of Ricci solitons with a physical interpretation of the notion of the vector field occurring in the definition. We investigate the geometrical symmetries of Petrov type N pure radiation fields along the vector field associated to Ricci solitons.

1. INTRODUCTION

It is known that Petrov type N solutions of Einstein vacuum equations are among the most interesting, rather difficult and little explored of all empty spacetime metrics ([18], [26]). From the physical point of view, they represent spacetime filled up entirely with gravitational radiation while mathematically they form a class of solutions of Einstein equations which should be possible to be determined explicitly. The behavior of the gravitational radiation from a bounded source is an important physical problem. Even reasonably far from the source, however, twisting type N solutions of the vacuum field equations are required for an exact description of that radiation. Such solutions would provide small laboratories in which to understand better the complete nature of singularities of type N solutions and could also be used to check numerical solutions that include gravitational radiation [28].

Moreover, in general theory of relativity the curvature tensor describing the gravitational field consists of two parts viz., the matter part and the free gravitational part. The interaction between these two parts is described through Bianchi identities. For a given

---

2000 *Mathematics Subject Classification.* 53C25, 53C80, 83C20.
Key words and phrases. Einstein spaces, type N fields, symmetries and Ricci solitons.

distribution of matter, the construction of gravitational potential satisfying Einstein's field equations is the principal aim of all investigations in gravitational physics and this has often been achieved by imposing symmetries on the geometry compatible with the dynamics of the chosen distribution of matter. The geometrical symmetries of the spacetime are expressible through the vanishing of the Lie derivative of certain tensors with respect to a vector.

In a series of papers ([13]-[15], [19]-[21]) Katzin, Levine, Davis and collaborators have identified a number of symmetries for the gravitational field with their interrelationships and have obtained the corresponding weak conservation laws as the integrals of the geodesic equation. Different types of matter distribution compatible with geometrical symmetries have been the subject of interest of several investigators for quite sometime and in this connection, Oliver and Davis [24], for the perfect fluid spacetimes, have studied the time-like symmetries with special reference to conformal motion and family of contracted Ricci collineation. The perfect fluid spacetimes including electromagnetic field which admit symmetry mapping belonging to the family of contracted Ricci collineation, have been studied by Norris et al. [23]. The role of geometrical symmetries in the study of fluid spacetimes, with an emphasis on conformal collineation has been explored by Duggal [16] and Duggal and Sharma [17] (see also [4]). The geometrical symmetry $\pounds_\xi R_{ij} = 2\Omega R_{ij}$, known as Ricci inheritance, has been studied by Ahsan [6], who obtained the necessary and sufficient conditions for perfect fluid spacetimes to admit such symmetries in terms of the kinematical quantities (also see [1]-[3]). Different types of symmetries of Petrov type N gravitational fields has been the subject of interest since last few decades (cf., [5]) but a complete analysis of collineations is not found in the literature (as far as we know).

Recently geometric flows have become important tools in Riemannian geometry and general relativity. List [10] has studied a geometric flow whose fixed points correspond to static Ricci flat spacetime which is nothing but Ricci flow pullback by a certain diffeomorphism. The association of each Ricci flat spacetime gives notion of local Ricci solitons in one higher dimension. The importance of geometric flow in Riemannian geometry is due to Hamilton who has given the flow equation and List generalized Hamilton's equation and extend it to spacetime for static metrics. He has given system of flow equations whose fixed points solve the Einstein free-scalar field system. This observation is useful for the correspondence of solutions of system i.e., Ricci solitons and symmetry properties of spacetime, that how Riemannian space (or spacetime) with Ricci solitons deals different kind of symmetry properties.

Motivated by the role of symmetries and Ricci solitons, a study of vector field involved in the definition of Ricci solitons and symmetries of spacetime is made. In section 2 preliminaries are given. The main results on the relationship between the symmetries of Petrov type N pure radiation fields and Ricci solitons has been given in section 3. Finally section 4 deals with the conclusion.

2. PRELIMINARIES

The geometrical notion known as Ricci solitons is shouldered on the concepts of Hamilton's flow. So far it has been discussed in different settings of Riemannian manifolds.

(a) Ricci Solitons A family $g_\lambda = g(\lambda; x)$ of Riemannian metrics on a n-dimensional ($n \geq 3$) smooth manifold M with parameter λ ranging in a time interval $J \subset \mathbb{R}$ including zero is called a Ricci flow if the Hamilton's equations

$$\frac{\partial g_0}{\partial \lambda} = -2Ric_0 \qquad (1)$$

of the Ricci flow (cf; [11], [12]) for $g_0 = g(0)$ and the Ricci tensor Ric_0 of the g_0 are satisfied. Corresponding to self similar solution of equation (1) is the notion of the local Ricci solitons, defined as a metric g_0 satisfying the equation

$$-2Ric_0 = \pounds_\xi g_0 + 2kg_0 \qquad (2)$$

for vector field ξ on V_n and a constant k. The Ricci solitons is said to be steady (static) if $k = 0$, shrinking if $k < 0$ and expanding if $k > 0$. The metric g_0 is called a gradient Ricci solitons if $\xi = \nabla \phi$ i.e., gradient of some function ϕ. For Schwarzschild metric, Akbar and Woolger [9] have derived the expressions around this notion; while Ali and Ahsan [8] have studied this concept for obtaining the Gaussian curvature of Schwarzschild solitons.

For n-dimensional Riemannian manifold equation (2) can be written in general as

$$R_{ij} - \frac{1}{2}\mathcal{L}_\xi g_{ij} = k g_{ij} \tag{3}$$

So far more than twenty seven different types of collineations have been studied and the literature on such collineations is very large and still expanding with results of elegance (cf., [6]). However, here we shall mention only those symmetry assumptions that are required for subsequent investigation and we have

(b) Motion A spacetime is said to admit motion if there exists a vector field ξ^a such that

$$\mathcal{L}_\xi g_{ij} = \xi_{i;j} + \xi_{j;i} = 0 \tag{4}$$

Equation (4) is known as Killing equation and vector ξ^a is called a Killing vector field (cf. [27]).

(c) Conformal Motion (Conf M) If

$$\mathcal{L}_\xi g_{ij} = \sigma g_{ij} \tag{5}$$

where σ is a scalar, then the spacetime is said to admit conformal motion and vector field ξ is called a conformal Killing vector field.

(d) Special Conformal Motion (SCM) A spacetime admits SCM if

$$\mathcal{L}_\xi g_{ij} = \sigma g_{ij}, \quad \sigma_{;ij} = 0 \tag{6}$$

(e) Curvature Collineation (CC) A spacetime admits curvature collineation if there is a vector field ξ^i such that

$$\mathcal{L}_\xi R^i_{jkl} = 0 \tag{7}$$

where R^i_{jkl} is Riemann curvature tensor.

(f) Ricci Collineation (RC) A spacetime is said to admit Ricci collineation if there is a vector field ξ^i such that

$$\mathcal{L}_\xi R_{ij} = 0 \tag{8}$$

where R_{ij} is the Ricci tensor.

(g) Affine Collineation (AC) If
$$\mathcal{L}_\xi \Gamma^i_{jk} = \xi^i_{;jk} + R^i_{jmk}\xi^m = 0 \tag{9}$$
then the spacetime is said to admit an AC.

(h) Weyl Projective Collineation (WPC)
A symmetry property of a spacetime is called WPC if and only if
$$\mathcal{L}_\xi W^i_{jkl} = 0 \ (n > 2) \tag{10}$$
where W^i_{jkl} is Weyl projective tensor.

3. Main Results

In this section, we shall discuss the role of Ricci solitons in the study of Einstein spaces and Petrov type N pure radiation fields. In 4-dimensional spacetime, the Weyl tensor is related to the Riemann and Ricci tensors through the equation

$$\begin{aligned} C_{ijkl} &= R_{ijkl} - \frac{1}{2}(g_{ik}R_{jl} + g_{jl}R_{ik} - g_{jk}R_{il} - g_{il}R_{jk}) \\ &+ \frac{1}{6}(g_{ik}g_{jl} - g_{il}g_{jk})R \end{aligned} \tag{11}$$

In NP-formalism (cf., [22]), the components of Weyl tensor are expressed by five complex scalars Ψ_0, Ψ_1, Ψ_2, Ψ_3 and Ψ_4. Through these components the gravitational field has been classified into six categories type I, II, D, III, N and O (cf. [26]). The Weyl scalar along with Goldberg-Sachs theorem declares type N pure radiation field follow the conditions

$$\Psi_4 = \Psi \neq 0, \ \Psi_i = 0, \ i = 0, 1, 2, 3 \tag{12}$$

and
$$\kappa = \sigma = \epsilon = 0 \tag{13}$$
where κ, σ, ϵ are the spin-coefficients [22]. Ali and Ahsan [7] have obtained symmetries for Weyl conformal tensor. Using equations (11)-(13) and definitions (b)-(c), we can write

Lemma 1 *In type N PR fields every conformal motion, special conformal motion and homothetic motion, all degenerate to motion.*

From equations (3) and (4), we have

$$2R_{ij} = \pounds_\xi g_{ij} + 2kg_{ij} \qquad (14)$$

$$= \xi_{i;j} + \xi_{j;i} + 2kg_{ij}$$

Contracting this equation with g^{ij}, we get

$$R = \xi^i_{;i} + kn \qquad (15)$$

which can be expressed as

$$div\xi = \nabla_i \xi^i = (R - kn) \qquad (16)$$

where $R = g^{ij}R_{ij}$ is scalar curvature. From equations (14) and (16), we get

$$(n^{-1}Rg_{ij} - R_{ij}) = -\frac{1}{2}\pounds_\xi g_{ij} + n^{-1}(div\xi)g_{ij} \qquad (17)$$

Now for g_{ij} to be Einstein metric i.e., $R_{ij} = \mu g_{ij}$ where μ can be chosen as $n^{-1}R$, equation (14) together with the definition of conformal motion ($\pounds_\xi g_{ij} = \sigma g_{ij}$), gives

Lemma 2 [25] *The vector field ξ associated with Ricci solitons (M, g) is conformally Killing if and only if (M, g) is an Einstein manifold of dimension $(n \geq 3)$.*

Using Lemmas 1 and 2, we can state

Theorem 1 *Type N pure radiation fields admit motion along a vector field ξ associated to Ricci solitons (M, g) if and only if M is an Einstein space.*

For Killing vector field ξ, equation (3) reduces to

$$R_{ij} = kg_{ij} \qquad (18)$$

Taking Lie derivative with respect to vector field ξ

$$\pounds_\xi R_{ij} = k\pounds_\xi g_{ij} = 0$$

Thus, we have

Theorem 2 *A vector field ξ associated to Ricci solitons (M, g) is Ricci collineation vector field in Type N PR fields if g is Einstein metric.*

Taking the Lie derivative of Christoffel symbol
$\Gamma^i_{jk} = \frac{1}{2}g^{il}\left(\frac{\partial g_{jl}}{\partial \xi^k} - \frac{\partial g_{jk}}{\partial \xi^l} + \frac{\partial g_{kl}}{\partial \xi^j}\right)$ along the vector field ξ, after a careful calculation we get

$$\mathcal{L}_\xi \Gamma^i_{jk} = \xi^i_{;jk} + R^i_{jmk}\xi^m \tag{19}$$

Now if ξ is Killing vector field, then

$$\xi^i_{;jk} + R^i_{jmk}\xi^m = 0 \tag{20}$$

where

$$R^h_{ijk} = -\frac{\partial \Gamma^h_{ij}}{\partial x^k} + \frac{\partial \Gamma^h_{ik}}{\partial x^j} - \Gamma^a_{ij}\Gamma^h_{ak} + \Gamma^b_{ik}\Gamma^h_{bj} \tag{21}$$

is the Reimann curvature tensor.

Using equations (19) and (20) along with the definition of affine collineation, we have

Theorem 3 *Type N pure radiation fields admit affine collineation along a Killing vector field ξ associated to Ricci solitons (M, g) if and only if M is an Einstein space.*

From the definition of Lie derivative

$$\mathcal{L}_\xi R^i_{jkl} = \xi^h R^i_{jkl;h} - R^h_{jkl}\xi^i_{;h} + R^i_{hkl}\xi^h_{;j} + R^i_{jhl}\xi^h_{;k} + R^i_{jkh}\xi^h_{;l} \tag{22}$$

while using the definition of Christoffel symbol and Killing vector field ξ, we have

$$\mathcal{L}_\xi R^i_{jkl} = 0$$

which establishes the curvature collineation, so we have

Theorem 4 *A Killing vector field ξ associated to Ricci solitons (M, g) is Curvature collineation vector field in type N PR fields if g is Einstein metric.*

The Weyl projective tensor is given by

$$W^i_{jkl} = R^i_{jkl} - \tfrac{1}{3}(R_{jk}\delta^i_j - R_{jl}\delta^i_k) \tag{23}$$

For $R_{ij} = 0$, $W^i_{jkl} = R^i_{jkl}$ or $W_{ijkl} = R_{ijkl}$

From equations (7) and (23), we can easily write

Lemma 3 [20] *In a Riemannian manifold curvature collineation implies the Weyl projective collineation but converse is true for empty spacetimes.*

So, theorem 4 and lemma 4 constitute the following

Corollary 1 *A Killing vector field ξ associated to Ricci solitons (M, g) is Weyl Projective collineation vector field in type N PR fields if g is Einstein metric.*

4. CONCLUSION

For Einstein spaces different kind of symmetry properties for type N pure radiation fields are established with the help of vector field associated with Ricci solitons. There are other symmetries for type N which can be obtained through the existence of Killing vectors corresponding to Ricci solitons. Further the results on geometrical symmetries can be obtained in plenty for spaces of constant curvature (space forms) because these are maximally symmetric spaces.

REFERENCES

[1] Ahsan, Z.: *"Collineation in electromagnetic field in general relativity- The null field case"*. Tamkang Journal of Maths. 9, No. 2 (1978) 237.
[2] Ahsan, Z.: *"On the Nijenhuis tensor for null electromagnetic field"*. Journal of Math. Phys. Sci., 21 No.5 (1987)515-526.
[3] Ahsan, Z.: *"Symmetries of the Electromagnetic Fields in General Relativity"*. Acta Phys. Sinica 4, (1995) 337.

[4] Ahsan, Z. : " *A symmetry property of the spacetime of general relativity in terms of the space matter tensor*" Brazilian Journal of Phys., 26 No.3 (1996) 572-576.
[5] Ahsan, Z.: *"Interacting radiation field"* Indian J. Pure App. Maths., 31 (2) (2000) 215-225.
[6] Ahsan, Z.: *"On a geometrical symmetry of the spacetime of genearal relativity"* Bull. Cal. Math. Soc., 97 (3) (2005) 191-200.
[7] Ali, M. and Ahsan, Z.: *"Ricci Solitons and Symmetries of Spacetime Manifold of General Relativity"*., Glob. J. Adv. Res. Class. Mod. Geom., 1, No. 2 (2013) 75-84.
[8] Ali, M. and Ahsan, Z.: *"Gravitational Field of Schwarzschild Soliton"*. Arab J. Math. Sci. http://dx.doi.org/10.1016/j.ajmsc.2013.10.003.
[9] Akbar, M.M. and Woolger, E.: *"Ricci soliton and Einstein-Scalar Field Theory"*., Class. Quantum Grav. 26 (2009) 55015.
[10] B List: *"Evolution of an extended Ricci flow system"* PhD thesis 2005.
[11] Chow, B., Lu, P. and Ni, L.: *"Hamilton's Ricci Flow"* Amer. Math. Soc. Province, RI, (2004).
[12] Chow, B. and Knopf, D.: *"The Ricci Flow; an Introduction"* Amer. Math. Soc. Province, RI, (2004).
[13] Davis, W.R., Green, L.H. and Norris, L.K. : *"Relativistic matter fields admitting Ricci collineations and elated conservation laws"* Il Nuovo Cimento, 34B (1976) 256-280.
[14] Davis, W.R. and Moss, M.K.: *"Conservation laws in the general theory of relativity"* Il Nuovo Cimento, 65B (1970) 19-32.
[15] Davis, W.R. and Oliver, Jr. Dr.: *"Matter field space times admitting symmetry mappings satisfying vanishing contraction of the Lie deformation of the Ricci tensor"*. Ann. Inst. Henri Poincare, 26, No.2 (1978) 197.
[16] Duggal, K.L.: *"Relativistic fluids with shear and timelike conformal collineations"* J. Math. Phys., 28 (1987) 2700-2705.
[17] Duggal, K.L. and Sharma R.: *"Conformal collineations and anisotropic fluids in general relativity* J. Math. Phys., 27 (1986) 2511-2514.
[18] Garcia Diaz, A. and Plebanski, J.F.: *"All nontwisting N's with cosmological constant"* J. Math. Phys. 22 (1981) 2655-2659.
[19] Katzin, G.H. and Levine, J.: *"Applications of Lie derivatives to symmetries, geodesic mappings, and first integrals in Riemannian spaces"* J. Colloq. Math., 26 (1972) 21-38.
[20] Katzin, G.H., Levine, J. and Davis, W.R.: *"Curvature Collineations: A Fundamental symmetry property of the space-times of general relativity defined by the vanishing Lie derivative of the Riemann curvature tensor"* J. Math. Phys., 10 (1969) 617-630.
[21] Katzin, G.H., Livine, J. and Davis, W.R.: *"Groups of curvature collineations in Riemannian space-times which admit fields of parallel vectors"*. J. Math. Phys., 11 (1970) 1578-1580.
[22] Newman, E.T. and Penrose, R.: *" An Approach to Gravitational Radiation by a Method of Spin Coefficients"* J. Math. Phys., 3 (1962) 566-579.

[23] Norris, L.K., Green, L.H. and Davis, W.R.: *"Fluid space-times including electromagnetic fields admitting symmetry mappings belonging to the family of contracted Ricci collineations"* J. Math. Phys., 18 (1977) 1305-1312.

[24] Oliver, D.R., and Davis, W.R.: *"On certain timelike symmetry properties and the evolution of matter field space-times that admit them"* Gen. Rel. Grav., 8 (1977) 905-914.

[25] Stepanov, S.E. and Shelepova, V.N.: *"A note on Ricci solitons"*. Mathematicheskie Zametici 86, No.3 (2009) 474-477.

[26] Stephani, H., Krammer, D., McCallum, M. and Herlt, E. : *"Exact solutions of Einsteins field equations"*, Cambridge University Press, Cambridge (2003).

[27] Yano, K.: *"The theory of Lie derivatives and its Application"* Vol. III, North Holand publishing co. Amsterdam p. Noordhoff L.T.D. Groningen (1957).

[28] Zakharov, V.D.: *"Gravitational waves in Einstein theory"*, Halsted Press, John Wiley & sons, New York (1973).

Department of Mathematics
Aligarh Muslim University, Aligarh-202 002, INDIA
Email: 1- zafar.ahsan@rediffmail.com, 2- musavvir.alig@gmail.com

Differential Geometry, Functional Analysis and Applications
Editors: Mohammad Hasan Shahid, Sharfuddin Ahmad et al.
Copyright © 2015, Narosa Publishing House, New Delhi

SEMI-SYMMETRIC LIGHTLIKE HYPERSURFACES OF AN INDEFINITE KENMOTSU SPACE FORM

RAM SHANKAR GUPTA

ABSTRACT. In this paper, we study semi-symmetric lightlike hypersurfaces of an indefinite Kenmotsu space form with structure vector field tangent to hypersurface. Also, I have given an example of totally geodesic semi-symmetric lightlike hypersurface in R_2^7.

1. INTRODUCTION

A semi-Riemannian manifold is called semi-symmetric if $R(X,Y) \cdot R = 0$, where $R(X,Y)$ is the curvature operator act as a derivative on R. It is well known that the class of semisymmetric manifolds includes the set of locally symmetric manifolds ($\nabla R = 0$) as a proper subset. Semisymmetric Riemannian manifolds were first studied by E. Cartan, A. Lichnerowicz, R.S. Couty and N.S. Sinjukov. In [3] K. Nomizu asked the question if there exist complete, irreducible and simply connected Riemannian manifolds of dimension $n \geq 3$ semi-symmetric and not locally symmetric. The first positive example was constructed in [6]. A general study of semi-symmetric Riemannian manifolds was made by Szabo [5].

In the theory of hypersurfaces of semi-Riemannian manifolds it is interesting to study the geometry of lightlike hypersurfaces due to the fact that the intersection of normal vector bundle and the tangent bundle is non-trivial. Thus, the study becomes more interesting and remarkably different from the study of non-degenerate hypersurfaces. The geometry of lightlike hypersurfaces of semi-Riemannian manifolds was studied in [1]. The lightlike hypersurfaces of semi-Euclidean spaces satisfying curvature conditions of semi-symmetry type was studied in [4].

2000 *Mathematics Subject Classification.* 53C15, 53C40, 53C50, 53D15.

Key words and phrases. Degenerate metric, lightlike hypersurfaces, indefinite Kenmotsu space form.

The purpose of the present paper is to study the semi-symmetric lightlike hypersurface of indefinite Kenmotsu space form with structure vector field ξ tangent to hypersurface.

In Section 2, I have collected the formulae and information which are useful in our subsequent sections. Section 3, is devoted to study the semi-symmetric lightlike hypersurfaces of an indefinite Kenmotsu space form. Also, I have given an example of totally geodesic semi-symmetric lightlike hypersurface in R_2^7.

2. Preliminaries

An odd-dimensional semi-Riemannian manifold $\overline{M}$ is said to be an indefinite almost contact metric manifold if there exist structure tensors $\{\phi, \xi, \eta, \overline{g}\}$, where ϕ is a (1,1) tensor field, ξ a vector field, η a 1-form and $\overline{g}$ is the semi-Riemannian metric on $\overline{M}$ satisfying
(2.1)
$$\begin{cases} \phi^2 X = -X + \eta(X)\xi, \quad \eta \circ \phi = 0, \quad \phi\xi = 0, \quad \eta(\xi) = 1 \\ \overline{g}(\phi X, \phi Y) = \overline{g}(X, Y) - \eta(X)\eta(Y), \quad \overline{g}(X, \xi) = \eta(X) \end{cases}$$

for any $X, Y \in \Gamma(T\overline{M})$, where $\Gamma(T\overline{M})$ denotes the Lie algebra of vector fields on $\overline{M}$.

An indefinite almost contact metric manifold $\overline{M}$ is called an indefinite Kenmotsu manifold if [2],

(2.2) $(\overline{\nabla}_X \phi)Y = g(\phi X, Y)\xi - \eta(Y)\phi X, \quad \text{and} \quad \overline{\nabla}_X \xi = X - \eta(X)\xi$

for any $X, Y \in T\overline{M}$, where $\overline{\nabla}$ denote the Levi-Civita connection on $\overline{M}$.

An indefinite almost contact metric manifold $\{\overline{M}, \phi, \xi, \eta, \overline{g}\}$ is called an indefinite Kenmotsu space form $\overline{M}(c)$ if it satisfies [2]

(2.3)
$$\begin{aligned} \overline{R}(X,Y)Z = {} & \tfrac{c-3}{4}\{g(Y,Z)X - g(X,Z)Y\} \\ & + \tfrac{c+1}{4}\{g(X,\phi Z)\phi Y - g(Y, \phi Z)\phi X \\ & + 2g(X, \phi Y)\phi Z + \eta(X)\eta(Z)Y - \eta(Y)\eta(Z)X \\ & + g(X,Z)\eta(Y)\xi - g(Y,Z)\eta(X)\xi\} \end{aligned}$$

for any $X, Y, Z \in \Gamma(T\overline{M})$.

We write as follows:

(2.4) $$\overline{R}(X,Y,Z,W) = \overline{g}(\overline{R}(X,Y)Z,W)$$

(2.5) $$Ric(X,Y) = trace\{Z \to \overline{R}(X,Z)Y\}$$

where Ric denotes the Ricci tensor on $\overline{M}$ for $X,Y,Z,W \in \Gamma(T\overline{M})$.

For a $(0,k)$-tensor field T on $\overline{M}$, $k \geq 1$, the $(0, k+2)$ tensor field $\overline{R} \cdot T = 0$ is called curvature conditions of semi-symmetry type [4] and given by

(2.6)
$$(\overline{R}.T)(X_1, ..., X_k, X, Y) = -T(\overline{R}(X,Y)X_1, X_2, ..., X_k)$$
$$-... - T(X_1, ..., X_{k-1}, \overline{R}(X,Y)X_k)$$

for $X, Y, X_1, X_k \in \Gamma(T\overline{M})$.

A semi-Riemannian space form $\overline{M}$ is said to be semi-symmetric if $\overline{R} \cdot \overline{R} = 0$. Thus, from (2.6) and properties of curvature tensor, we have

(2.7)
$$(\overline{R}(X,Y).\overline{R})(U,V)W = \overline{R}(X,Y)\overline{R}(U,V)W - \overline{R}(U,V)\overline{R}(X,Y)W$$
$$-\overline{R}(\overline{R}(X,Y)U,V)W - \overline{R}(U, \overline{R}(X,Y)V)W = 0,$$

for any $X, Y, U, V, W \in \Gamma(T\overline{M})$.

Let (M,g) be a hypersurface of a $(2m+1)$-dimensional semi-Riemannian manifold $(\overline{M}, \overline{g})$ with index s, $0 < s < 2m+1$ and $g = \overline{g}_{|M}$. Then M is lightlike hypersurface of $\overline{M}$ if g is of constant rank $(2m-1)$ and the normal bundle $TM^\perp$ is a distribution of rank 1 on M [1]. A non-degenerate complementary distribution $S(TM)$ of rank $(2m-1)$ to $TM^\perp$ in TM, that is, $TM = TM^\perp \perp S(TM)$, is called screen distribution. The following result (cf. [1], Theorem 1.1, page 79) has an important role in studying the geometry of lightlike hypersurface.

Theorem A. Let $(M, g, S(TM))$ be a lightlike hypersurface of $\overline{M}$. Then, there exists a unique vector bundle $tr(TM)$ of rank 1 over M such that for any non-zero section E of $TM^\perp$ on a coordinate neighbourhood $U \subset M$, there exists a unique section N of $tr(TM)$

on U satisfying $\overline{g}(N,E) = 1$ and $\overline{g}(N,N) = \overline{g}(N,W) = 0$, $\forall W \in \Gamma(S(TM)|_u)$.

Then, we have the following decomposition:
(2.8)
$$TM = S(TM) \perp TM^\perp, \qquad \overline{TM} = S(TM) \perp (TM^\perp \oplus tr(TM)).$$

Throughout this paper, all manifolds are supposed to be paracompact and smooth. We denote by $\Gamma(E)$ the smooth sections of the vector bundle E, by $\perp$ and $\oplus$ the orthogonal and the non-orthogonal direct sum of two vector bundles, respectively.

Let $\overline{\nabla}$, ∇ and ∇^t denote the linear connections on $\overline{M}$, M and vector bundle $tr(TM)$, respectively. Then, the Gauss and Weingarten formulae are given by

(2.9) $\qquad \overline{\nabla}_X Y = \nabla_X Y + h(X,Y)$, $\forall X, Y \in \Gamma(TM)$

(2.10) $\qquad \overline{\nabla}_X V = -A_V X + \nabla^t_X V$, $\forall V \in \Gamma(tr(TM))$

where $\{\nabla_X Y, A_V X\}$ and $\{h(X,Y), \nabla^t_X V\}$ belongs to $\Gamma(TM)$ and $\Gamma(tr(TM))$,

respectively and A_V is the shape operator of M with respect to V. Moreover, in view of decomposition (2.9), equations (2.10) and (2.11) take the form

(2.11) $\qquad \overline{\nabla}_X Y = \nabla_X Y + B(X,Y)N$

(2.12) $\qquad \overline{\nabla}_X N = -A_N X + \tau(X)N$

for any $X, Y \in \Gamma(TM)$ and $N \in \Gamma(tr(TM))$, where $B(X,Y)$ and $\tau(X)$ are local second fundamental form and a 1-form on U, respectively. It follows that

$$B(X,Y) = \overline{g}(\overline{\nabla}_X Y, E) = \overline{g}(h(X,Y), E), B(X,E) = 0, \text{ and}$$
$$\tau(X) = \overline{g}(\nabla^t_X N, E).$$

Let P denote the projection morphism of $\Gamma(TM)$ on $\Gamma(S(TM))$ and ∇^*, ∇^{*t} denote the linear connections on $S(TM)$ and $STM^\perp$, respectively. Then from the decomposition of tangent bundle of lightlike hypersurface, we have

(2.13) $\qquad \nabla_X PY = \nabla^*_X PY + h^*(X, PY)$

(2.14) $\qquad \nabla_X E = -A^*_E X + \nabla^{*t}_X E$

for any $X, Y \in \Gamma(TM)$ and $E \in \Gamma(TM^\perp)$, where h^*, A^* are the second fundamental form and the shape operator of distribution $S(TM)$ respectively.

By direct calculations using Gauss-Weingarten formulae, (2.14) and (2.15), we find

(2.15) $\quad g(A_N Y, PW) = \bar{g}(N, h^*(Y, PW)); \quad\quad \bar{g}(A_N Y, N) = 0,$

(2.16) $\quad g(A_E^* X, PY) = \bar{g}(E, h(X, PY)); \quad\quad \bar{g}(A_E^* X, N) = 0,$

for any $X, Y, W \in \Gamma(TM)$, $E \in \Gamma(TM^\perp)$ and $N \in \Gamma(tr(TM))$.

Locally, we define on U

(2.17) $\quad C(X, PY) = \bar{g}(h^*(X, PY), N), \quad\quad \lambda(X) = \bar{g}(\nabla_X^{*t} E, N).$

Hence,

(2.18) $\quad h^*(X, PY) = C(X, PY)E, \quad\quad \nabla_X^{*t} E = \lambda(X)E.$

On the other hand, by using (2.12), (2.13), (2.15) and (2.18), we obtain

$$\lambda(X) = \bar{g}(\nabla_X E, N) = \bar{g}(\overline{\nabla}_X E, N) = -\bar{g}(E, \overline{\nabla}_X N) = -\tau(X).$$

Thus, locally (2.14) and (2.15) become

(2.19)
$\nabla_X PY = \nabla_X^* PY + C(X, PY)E, \quad\quad \nabla_X E = -A_E^* X - \tau(X)E.$

Finally, (2.16) and (2.17), locally become

(2.20) $\quad g(A_N Y, PW) = C(Y, PW); \quad\quad \bar{g}(A_N Y, N) = 0,$

(2.21) $\quad g(A_E^* X, PY) = B(X, PY); \quad\quad \bar{g}(A_E^* X, N) = 0.$

We note that second equation of (2.21) implies that $A_N X \in \Gamma(S(TM))$ for
$X \in \Gamma(TM)$, i.e. A_N is $\Gamma(S(TM))$ valued. On the other hand, from $\bar{g}(\overline{\nabla}_X E, E) = 0$, we have

(2.22) $\quad\quad\quad\quad\quad B(X, E) = 0.$

In general, the induced connection ∇ on M is not a metric connection. Since $\overline{\nabla}$ is a metric connection, we have

$$0 = (\overline{\nabla}_X \bar{g})(Y, Z) = X(\bar{g}(Y, Z)) - \bar{g}(\overline{\nabla}_X Y, Z) - \bar{g}(Y, \overline{\nabla}_X Z).$$

By using (2.12) in this equation, we obtain
(2.23)
$$(\nabla_X g)(Y,Z) = B(X,Y)\theta(Z) + B(X,Z)\theta(Y), \quad X,Y \in \Gamma(S(TM)|_u),$$
where θ is a differential 1-form locally defined on M by $\theta(\cdot) = \overline{g}(N,\cdot)$.

If $\overline{R}$ and R are the curvature tensors of $\overline{M}$ and M, then using (2.12) in the equation $\overline{R}(X,Y)Z = \overline{\nabla}_X \overline{\nabla}_Y Z - \overline{\nabla}_Y \overline{\nabla}_X Z - \overline{\nabla}_{[X,Y]} Z$, we obtain
(2.24)
$$\overline{R}(X,Y)Z = R(X,Y)Z + B(X,Z)A_N Y - B(Y,Z)A_N X$$
$$+ \{(\nabla_X B)(Y,Z) - (\nabla_Y B)(X,Z) + \tau(X)B(Y,Z) - \tau(Y)B(X,Z)\}N$$

(2.25) $\quad (\nabla_X B)(Y,Z) = XB(Y,Z) - B(\nabla_X Y, Z) - B(Y, \nabla_X Z).$

3. Semi-symmetric Lightlike Hypersurfaces in Indefinite Kenmotsu Space Form

In this section, we consider semi-symmetric lightlike hypersurfaces M in an indefinite Kenmotsu space form $\overline{M}(c)$.

For $X \in \Gamma(TM)$, we write
(3.1)
$$\phi X = tX + \beta(X)N$$
where tX is the tangential parts of ϕX and β is the one form on M.

Definition B. Let M be a lightlike hypersurface of a $(2m+1)$-dimensional indefinite Kenmotsu space form $\overline{M}(c)$. We say that M is semi-symmetric if the following condition is satisfied
(3.2)
$$(R(X,Y) \cdot R)(X_1, X_2, X_3, X_4) = 0$$
for $X, Y, X_1, X_2, X_3, X_4 \in \Gamma(TM)$.
We note that $(R(X,Y) \cdot R)(X_1, X_2, X_3, E) = 0$ for $E \in \Gamma(TM^\perp)$, therefore equation (3.2) reduces to
(3.3)
$$(R(X,Y) \cdot R)(X_1, X_2, X_3, PX_4) = 0.$$

We have following :
Lemma 3.1. Let M be a lightlike hypersurface of a $(2m+1)$-dimensional

indefinite Kenmotsu space form $\overline{M}(c)$. Then the Gauss equation of M is given by
(3.4)
$$R(X,Y)Z = B(Y,Z)A_N X - B(X,Z)A_N Y +$$
$$\frac{c-3}{4}\{g(Y,Z)X - g(X,Z)Y\}$$
$$+\frac{c+1}{4}\{g(X,\phi Z)tY - g(Y,\phi Z)tX + 2g(X,\phi Y)tZ + \eta(X)\eta(Z)Y$$
$$-\eta(Y)\eta(Z)X + g(X,Z)\eta(Y)\xi - g(Y,Z)\eta(X)\xi\}.$$

Proof: From (2.3), (2.25), (3.1) and comparing the tangential part, we obtain (3.4).

Theorem 3.1. Let M be a totally geodesic lightlike hypersurface of $(2m+1)$-dimensional indefinite Kenmotsu space form $\overline{M}(c)$. Then, M is semi-symmetric if $c = -1$.

Proof: Let M be a lightlike hypersurface of indefinite Kenmotsu space form. Then, we have

(3.5)
$$g((R(X,Y).R)(U,V)W,Z) = g(R(X,Y).R(U,V)W,Z)$$
$$-g(R(U,V)R(X,Y)W,Z) - g(R(R(U,V)X,Y)W,Z)$$
$$-g(R(X,R(U,V)Y)W,Z)$$

$\forall\ X,Y,Z,U,V,W \in \Gamma(TM)$.

Using (3.4) and Definition B in (3.10), we obtain

$g((R(X,Y).R)(U,V)W,PZ)$
$= B(Y,R(U,V)W)g(A_N X,PZ) - B(X,R(U,V)W)g(A_N Y,PZ) -$
$B(V,R(X,Y)W)g(A_N U,PZ) + B(U,R(X,Y)W)g(A_N V,PZ) -$
$B(Y,W)g(A_N R(U,V)X,PZ) + B(R(U,V)X,W)g(A_N Y,PZ) -$
$B(R(U,V)Y,W)g(A_N X,PZ) + B(X,W)g(A_N R(U,V)Y,PZ) +$
$\frac{c-3}{4}\{g(Y,R(U,V)W)g(X,PZ) - g(X,R(U,V)W)g(Y,PZ) -$
$g(V,R(X,Y)W)g(U,PZ) + g(U,R(X,Y)W)g(V,PZ) -$
$g(Y,W)g(R(U,V)X,PZ) + g(R(U,V)X,W)g(Y,PZ)$
$- g(R(U,V)Y,W)g(X,PZ) + g(X,W)g(R(U,V)Y,PZ)\} +$

$\frac{c+1}{4}\{g(X,\phi R(U,V)W)g(tY,PZ) - g(Y,\phi R(U,V)W)g(tX,PZ) +$
$2g(X,\phi Y)g(tR(U,V)W,PZ) - g(U,\phi R(X,Y)W)g(tV,PZ) +$
$g(V,\phi R(X,Y)W)g(tU,PZ) - 2g(U,\phi V)g(tR(X,Y)W,PZ) -$
$g(R(U,V)X,\phi W)g(tY,PZ) + g(tR(U,V)X,PZ)g(Y,\phi W) -$
$2g(R(U,V)X,\phi Y)g(tW,PZ) - g(X,\phi W)g(tR(U,V)W,PZ) +$
$g(tX,PZ)g(R(U,V)Y,\phi W) - 2g(X,\phi R(U,V)Y)g(tW,PZ) +$
$\eta(X)\eta(R(U,V)W)g(Y,PZ) - \eta(Y)\eta(R(U,V)W)g(X,PZ) +$
$g(X,R(U,V)W)\eta(Y)\eta(PZ) - g(Y,R(U,V)W)\eta(X)\eta(PZ) -$
$\eta(U)\eta(R(X,Y)W)g(V,PZ) + \eta(V)\eta(R(X,Y)W)g(U,PZ) -$
$g(U,R(X,Y)W)\eta(V)\eta(PZ) + g(V,R(X,Y)W)\eta(U)\eta(PZ) -$
$\eta(R(U,V)X)\eta(W)g(Y,Z) + \eta(Y)\eta(W)g(R(U,V)X,PZ) -$
$g(R(U,V)X,W)\eta(Y)\eta(PZ) + g(Y,W)\eta(R(U,V)X)\eta(PZ) -$
$g(R(U,V)Y,PZ)\eta(X)\eta(W) + g(X,PZ)\eta(R(U,V)Y)\eta(W) -$
$g(X,W)\eta(R(U,V)Y)\eta(PZ) + g(R(U,V)Y,W)\eta(X)\eta(PZ)\}.$

Or,

$g((R(X,Y).R)(U,V)W,PZ)$

$= g(A_N X,PZ)[B(Y,A_N U)B(V,W) - B(Y,A_N V)B(U,W) +$
$\frac{c-3}{4}\{g(V,W)B(Y,U) - g(U,W)B(Y,V)\} +$
$\frac{c+1}{4}\{\eta(U)\eta(W)B(Y,V) - g(V,\phi W)B(Y,tU) + 2g(U,\phi V)B(Y,tW) +$
$\eta(U)\eta(W)B(Y,V) - \eta(V)\eta(W)B(U,Y) + g(U,W)\eta(V)B(Y,\xi) -$
$g(V,W)\eta(U)B(Y,\xi)\}] - g(A_N Y,PZ)[B(X,A_N U)B(V,W) -$
$B(X,A_N V)B(U,W) + \frac{c-3}{4}\{g(V,W)B(X,U) - g(U,W)B(X,V)\} +$
$\frac{c+1}{4}\{\eta(U)\eta(W)B(X,V) - g(V,\phi W)B(X,tU) +$
$2g(U,\phi V)B(X,tW) + \eta(U)\eta(W)B(X,V) - \eta(V)\eta(W)B(U,X) +$
$g(U,W)\eta(V)B(X,\xi) - g(V,W)\eta(U)B(X,\xi)\}] -$
$g(A_N U,PZ)[B(V,A_N X)B(Y,W) - B(V,A_N Y)B(X,W) +$
$\frac{c-3}{4}\{g(Y,W)B(V,X) - g(X,W)B(V,Y)\} +$
$\frac{c+1}{4}\{g(X,\phi W)B(V,tY) - g(Y,\phi W)B(V,tX) + 2g(X,\phi Y)B(V,tZ) +$
$\eta(X)\eta(W)B(V,Y) - \eta(Y)\eta(W)B(V,X) + g(X,W)\eta(Y)B(V,\xi) -$
$g(Y,W)\eta(X)B(V,\xi)\}] + g(A_N V,PZ)[B(U,A_N X)B(Y,W) -$
$B(U,A_N Y)B(X,W) + \frac{c-3}{4}\{g(Y,W)B(U,X) - g(X,W)B(U,Y)\} +$
$\frac{c+1}{4}\{g(X,\phi W)B(U,tY) - g(Y,\phi W)B(U,tX) + 2g(X,\phi Y)B(U,tZ) +$
$\eta(X)\eta(W)B(U,Y) - \eta(Y)\eta(W)B(U,X) + g(X,W)\eta(Y)B(U,\xi) -$
$g(Y,W)\eta(X)B(U,\xi)\}] - B(Y,W)[g(A_N B(V,X)A_N U,PZ) -$
$g(A_N B(U,X)A_N V,PZ) + g(A_N \frac{c-3}{4}\{g(V,X)U - g(U,X)V\},PZ) +$
$g(A_N \frac{c+1}{4}\{g(U,\phi X)tV - g(V,\phi X)tU + 2g(U,\phi V)tX\},PZ) +$
$g(A_N \frac{c+1}{4}\{\eta(U)\eta(X)V - \eta(V)\eta(X)U + g(U,X)\eta(V)\xi -$

$$g(V,X)\eta(U)\xi\}, PZ)] + g(A_N Y, PZ)[B(V,X)B(A_N U, W) -$$
$$B(U,X)B(A_N V, W) + \tfrac{c-3}{4}\{g(V,X)B(U,W) - g(U,X)B(U,W)\} +$$
$$\tfrac{c+1}{4}\{g(U,\phi X)B(tV,W) - g(V,\phi X)B(tU,W) +$$
$$2g(U,\phi V)B(tX,W) + \eta(X)\eta(U)B(V,W) - \eta(X)\eta(V)B(U,W) +$$
$$g(X,U)\eta(V)B(\xi,W) - g(X,V)\eta(U)B(\xi,W)\}] -$$
$$g(A_N X, PZ)[B(U,Y)B(A_N U, W) - B(U,Y)B(A_N V, W) +$$
$$\tfrac{c-3}{4}\{g(V,Y)B(U,W) - g(U,Y)B(V,W)\} + \tfrac{c+1}{4}\{g(U,\phi Y)B(tV,W) -$$
$$g(V,\phi Y)B(tU,W) + 2g(U,\phi V)B(tY,W) + \eta(U)\eta(V)B(V,W) -$$
$$\eta(V)\eta(Y)B(U,W) + g(U,Y)\eta(V)B(\xi,W) - g(V,Y)\eta(U)B(\xi,W)\}] +$$
$$B(X,W)[g(A_N B(V,Y)A_N U, PZ) - g(A_N B(U,Y)A_N V, PZ) +$$
$$g(A_N \tfrac{c-3}{4}\{g(V,Y)U - g(U,Y)V\}, PZ) + g(A_N \tfrac{c+1}{4}\{g(U,\phi Y)tV -$$
$$g(V,\phi Y)tU + 2g(U,\phi V)tY\}, PZ) + g(A_N \tfrac{c+1}{4}\{\eta(U)\eta(Y)V -$$
$$\eta(V)\eta(Y)U + g(U,Y)\eta(V)\xi - g(V,Y)\eta(U)\xi\}, PZ) +$$
$$\tfrac{c-3}{4}\{g(Y,R(U,V)W)g(X,PZ) - g(X,R(U,V)W)g(Y,PZ) -$$
$$g(V,R(X,Y)W)g(U,PZ) + g(U,R(X,Y)W)g(V,PZ) +$$
$$g(R(U,V)X,W)g(Y,PZ) - g(R(U,V)X,PZ)g(Y,W) -$$
$$g(PZ,R(U,V)Y)g(X,W) - g(W,R(U,V)Y)g(X,PZ)\} +$$
$$\tfrac{c+1}{4}\{g(X,\phi R(U,V)W)g(tY,PZ) - g(Y,\phi R(U,V)W)g(tX,PZ) +$$
$$2g(X,\phi Y)g(PZ,tR(U,V)W) + g(V,\phi R(X,Y)W)g(tU,PZ) -$$
$$g(U,\phi R(X,Y)W)g(tV,PZ) - 2g(U,\phi V)g(PZ,tR(X,Y)W) -$$
$$g(R(U,V)X,\phi W)g(tY,PZ) + g(Y,\phi W)g(tR(U,V)X,PZ) -$$
$$2g(R(U,V)X,\phi Y)g(tW,PZ) - g(X,\phi W)g(tR(U,V)W,PZ) -$$
$$g(R(U,V)Y,\phi W)g(tX,PZ) - 2g(X,\phi R(U,V)Y)g(tW,PZ) +$$
$$\eta(X)\eta(R(U,V)W)g(Y,PZ) - \eta(Y)\eta(R(U,V)W)g(X,PZ) +$$
$$\eta(Y)\eta(PZ)g(R(U,V)W,X) - \eta(X)\eta(PZ)g(R(U,V)W,Y) -$$
$$g(V,PZ)\eta(U)\eta(R(X,Y)W) + g(U,PZ)\eta(V)\eta(R(X,Y)W) -$$
$$g(U,R(X,Y)W)\eta(V)\eta(PZ) + g(V,R(X,Y)W)\eta(U)\eta(PZ) -$$
$$\eta(R(U,V)X)\eta(W)g(Y,PZ) + \eta(Y)\eta(W)g((R(U,V)X,PZ) -$$
$$g(R(U,V)X,W)\eta(Y)\eta(PZ) + g(Y,W)\eta(R(U,V)X)\eta(PZ) -$$
$$g(R(U,V)Y,PZ)\eta(X)\eta(W) + g(X,PZ)\eta(R(U,V)Y)\eta(W) -$$
$$g(X,W)\eta(R(U,V)Y)\eta(PZ) + g(R(U,V)Y,W)\eta(X)\eta(PZ)\}.$$

Putting $Y = U = E \in \Gamma(TM^\perp)$ in above equation and a straight forward
calculations, we have

(3.6) $\quad g((R(X,E).R)(E,V)W, PZ) =$
$$-g(A_N E, PZ)[B(V,W)B(X,A_N E) - \tfrac{c+1}{4}\{g(V,\phi W)B(X,tE) -$$
$$2g(E,\phi V)B(X,tW)\} - B(V,A_N E)B(X,W) +$$

$$\frac{c+1}{4}\{g(X,\phi W)B(V,tE) - g(E,\phi W)B(V,tX) +$$
$$2g(X,\phi E)B(V,tZ)\} - B(V,X)B(A_N E, W) -$$
$$\frac{c+1}{4}\{g(E,\phi X)B(tV,W) -$$
$$g(V,\phi X)B(tE,W)2g(E,\phi V)B(tX,W)\}] +$$
$$g(A_N X, PZ)[\frac{c+1}{4}\{g(V,\phi E)B(tE,W) - 2g(E,\phi V)B(tE,W)\}] +$$
$$B(X,W)[g(A_N \frac{c+1}{4}\{-g(V,\phi E)B(tE,W) + 2g(E,\phi V)tE\}, PZ)] +$$
$$\frac{c-3}{4}[g(E, R(E,V)W)g(X, PZ) + g(E, R(X,E)W)g(V, PZ) +$$
$$g(X,W)g(R(E,V)E, PZ) - g(R(E,V)E, W)g(X, PZ)\} +$$
$$\frac{c+1}{4}\{g(X,\phi R(E,V)W)g(tE, PZ) - g(E,\phi R(E,V)W)g(tX, PZ) +$$
$$2g(X,\phi E)g(PZ, tR(E,V)W) + g(V,\phi R(X,E)W)g(tE, PZ) -$$
$$g(E,\phi R(X,E)W)g(tV, PZ) - 2g(E,\phi V)g(PZ, tR(X,E)W) -$$
$$g(R(E,V)X, \phi W)g(tE, PZ) + g(E,\phi W)g(tR(E,V)X, PZ) -$$
$$2g(R(E,V)X, \phi E)g(tW, PZ) - g(X,\phi W)g(tR(E,V)W, PZ) -$$
$$g(R(E,V)E, \phi W)g(tX, PZ) - 2g(X,\phi R(E,V)E)g(tW, PZ) +$$
$$\eta(X)\eta(R(E,V)W)g(E, PZ) - \eta(X)\eta(PZ)g(R(E,V)W, E) -$$
$$g(E, R(X,E)W)\eta(V)\eta(PZ) - g(R(E,V)E, PZ)\eta(X)\eta(W) +$$
$$g(X, PZ)\eta(R(E,V)E)\eta(W) - g(X,W)\eta(R(E,V)E)\eta(PZ) +$$
$$g(R(E,V)E, W)\eta(X)\eta(PZ)\}.$$

Taking $c = -1$ and using the fact that M is totally geodesic in above equation, we find

$$g((R(X,Y).R)(U,V)W, PZ) = 0,$$

which proves the theorem.

In the following example, $(R_q^{2m+1}, \overline{\phi}, \xi, \eta, \overline{g})$ will denote the manifold R_q^{2m+1} with its usual Kenmotsu structure given by

$$\eta = dz, \qquad \xi = \partial z,$$

$\overline{g} = \eta \otimes \eta - e^{2z}\{\sum_{i=1}^{q/2}(dx^i \otimes dx^i + dy^i \otimes dy^i)\} + e^{2z}\{\sum_{i=q}^{m} dx^i \otimes (dx^i + dy^i \otimes dy^i)\}$,

$\overline{\phi}(\sum_{i=1}^{m}(X_i \partial x^i + Y_i \partial y^i) + Z \partial z) = \sum_{i=1}^{m}(Y_i \partial x^i - X_i \partial y^i)$,

where (x^i, y^i, z) are the Cartesian coordinates.

Example 1. Let $\overline{M} = (R_2^7, \overline{g})$ be a semi-Euclidean space, where $\overline{g}$ is

of signature
$(-, +, +, -, +, +, +)$ with respect to the canonical basis
$$\{\partial x_1, \partial x_2, \partial x_3, \partial y_1, \partial y_2, \partial y_3, \partial z\}.$$
Consider a hypersurface M of R_2^7, defined by
$$X(u, v, \theta_1, \theta_2, s, t) = (u, u, v, \theta_1, \theta_2, s, t).$$
Then a local frame of TM is given by
$$Z_1 = e^{-z}\{\partial x_1 + \partial x_2\}, \ Z_2 = e^{-z}\partial x_3, \ Z_3 = e^{-z}\partial y_1, \ Z_4 = e^{-z}\partial y_2, \ Z_5 = e^{-z}\partial y_3, \ Z_6 = \xi = \partial z.$$
Hence, $TM^\perp = \text{span}\{Z_1\}$ and $tr(TM)$ is spanned by $N = \frac{e^{-z}}{2}(-\partial x_1 + \partial x_2)$. Using Gauss and Weingarten formulae, we obtain that

$$h(Z_i, Z_j) = 0 \quad \text{and} \quad \overline{\nabla}_{Z_i} N = 0, \quad \text{for } i, j = 1, ...6.$$

Hence M is totally geodesic semi-symmetric lightlike hypersurface, which support Theorem 3.1.

References

[1] Duggal, K. L. and Bejancu, A.: Lightlike Submanifolds of Semi-Riemannian Manifolds and Applications, Kluwer Acad. Publishers, Dordrecht, 1996.

[2] Kenmotsu, K.: A class of almost contact Riemannian manifolds. Tohoku Math. J. 21 (1972), 93103.

[3] Nomizu, K.: On hypersurfaces satisfying a certain condition on the curvature tensor, Tohoku Math. J. 20, 46-59, (1986).

[4] Sahin, B.: Lightlike hypersurfaces of semi-Euclidean spaces satisfying curvature conditions of semi-symmetry type, Turk. J. Math. 31, 139-162, (2007).

[5] Szabo, Z.: Structure theorems on Riemannian spaces satisfying $R(X, Y).R = 0$, the local version, J. Differential Geometry, 17, 531-582, (1982).

[6] Takagi, H.: An example of Riemannian manifolds satisfying $R(X, Y).R = 0$ but not $\nabla R = 0$, Tohoku Math. J. 24, 105-108, (1972)

Author's address:
Ram Shankar Gupta
University School of Basic and Applied Sciences,
Guru Gobind Singh Indraprastha University,
Dwarka Sec-16C, New Delhi-110075, India
Email: ramshankar.gupta@gmail.com

Differential Geometry, Functional Analysis and Applications
Editors: Mohammad Hasan Shahid, Sharfuddin Ahmad *et al.*
Copyright © 2015, Narosa Publishing House, New Delhi

GENERIC LIGHTLIKE SUBMANIFOLDS OF INDEFINITE KAEHLER MANIFOLDS

MANISH GOGNA, SANGEET KUMAR, RAKESH KUMAR AND R. K. NAGAICH

ABSTRACT. We introduce generic lightlike submanifolds of an indefinite Kaehler manifold and study the existence of this class in an indefinite complex space form. We find the conditions for the integrability of various distributions of a generic lightlike submanifold and also obtain conditions for the distributions to define totally geodesic foliation in generic lightlike submanifold. Finally we obtain necessary and sufficient condition for an induced connection to be a metric connection.

1. INTRODUCTION

In the geometry of lightlike submanifolds, the normal vector bundle intersects with the tangent bundle. Thus the study of lightlike submanifolds becomes more difficult and striking different from study of non-degenerate submanifolds, that is, one cannot use fundamental concepts of classical theory to define any induced objects on a lightlike submanifold,(for detail see [2]). Yano and Kon [6, 7] introduced generic submanifolds with positive definite metric. Therefore this geometry may not be applicable to the other branches of mathematics and physics, where the metric is not necessarily definite. Therefore the lightlike notion of generic submanifolds of indefinite Sasakian and indefinite cosymplectic manifolds is introduced recently in [3, 4]. The general notion of generic lightlike submanifolds of indefinite Keahler has not been introduced yet. Since the geometry of lightlike submanifolds is used in mathematical physics, in particular, in general relativity. Therefore we introduce generic lightlike submanifold of indefinite Kaehler manifolds and prove its

---

2000 *Mathematics Subject Classification.* 53C15, 53C40, 53C50.
Key words and phrases. Indefinite Kaehler manifold, Genric lightlike submanifold, integrability of distributions, metric connection.

existence in an indefinite complex space form. We find the conditions for the integrability of the various distributions of generic lightlike submanifolds of indefinite Kaehler manifold. We also find the conditions for the distributions D and D' to define totally geodesic foliation in generic lightlike submanifold. Finally we characterize the induced connection to be a metric connection on the genric lightlike submanifold.

2. Lightlike Submanifolds

Let $(\bar{M}, \bar{g})$ be a real $(m+n)$-dimensional semi-Riemannian manifold of constant index q such that $m, n \geq 1$, $1 \leq q \leq m+n-1$ and (M, g) be an m-dimensional submanifold of $\bar{M}$ and g the induced metric of $\bar{g}$ on M. If $\bar{g}$ is degenerate on the tangent bundle TM of M then M is called a lightlike submanifold of $\bar{M}$. For a degenerate metric g on M, $TM^\perp$ is a degenerate n-dimensional subspace of $T_x\bar{M}$. Thus both T_xM and $T_xM^\perp$ are degenerate orthogonal subspaces but no longer complementary. In this case, there exists a subspace $RadT_xM = T_xM \cap T_xM^\perp$ which is known as radical (null) subspace. If the mapping $RadTM : x \in M \longrightarrow RadT_xM$, defines a smooth distribution on M of rank $r > 0$ then the submanifold M of $\bar{M}$ is called an r-lightlike submanifold and $RadTM$ is called the radical distribution on M.

Screen distribution $S(TM)$ is a semi-Riemannian complementary distribution of $Rad(TM)$ in TM, that is, $TM = RadTM \perp S(TM)$ and $S(TM^\perp)$ is a complementary vector subbundle to $RadTM$ in $TM^\perp$. Let $tr(TM)$ and $ltr(TM)$ be complementary (but not orthogonal) vector bundles to TM in $T\bar{M} \mid_M$ and to $RadTM$ in $S(TM^\perp)^\perp$ respectively. Then we have

(2.1) $$tr(TM) = ltr(TM) \perp S(TM^\perp).$$

(2.2)
$$T\bar{M}\mid_M = TM \oplus tr(TM) = (RadTM \oplus ltr(TM)) \perp S(TM) \perp S(TM^\perp).$$

Let u be a local coordinate neighborhood of M and consider the local quasi-orthonormal fields of frames of $\bar{M}$ along M, on u as $\{\xi_1, ..., \xi_r, W_{r+1}, ..., W_n, N_1, ..., N_r, X_{r+1}, ..., X_m\}$, where $\{\xi_1, ..., \xi_r\}$, $\{N_1, ..., N_r\}$ are local lightlike bases of $\Gamma(RadTM\mid_u)$, $\Gamma(ltr(TM)\mid_u$) and $\{W_{r+1}, ..., W_n\}, \{X_{r+1}, ..., X_m\}$ are local orthonormal bases of $\Gamma(S(TM^\perp)\,|_u)$ and $\Gamma(S(TM)\,|_u)$ respectively. For this quasi-orthonormal fields of frames, we have

Theorem 2.1. *([2]). Let $(M, g, S(TM), S(TM^\perp))$ be an r-lightlike submanifold of a semi-Riemannian manifold $(\bar{M}, \bar{g})$. Then there exists a complementary vector bundle $ltr(TM)$ of $RadTM$ in $S(TM^\perp)^\perp$ and a basis of $\Gamma(ltr(TM)\,|_u)$ consisting of smooth section $\{N_i\}$ of $S(TM^\perp)^\perp\,|_u$, where u is a coordinate neighborhood of M such that*

$$\bar{g}(N_i, \xi_j) = \delta_{ij}, \quad \bar{g}(N_i, N_j) = 0, \text{for any } i,j \in \{1,2,..,r\},$$

where $\{\xi_1, ..., \xi_r\}$ is a lightlike basis of $\Gamma(Rad(TM))$.

Let $\bar{\nabla}$ be the Levi-Civita connection on $\bar{M}$ then according to the decomposition (2.2), the Gauss and Weingarten formulas are given by

(2.3) $\quad \bar{\nabla}_X Y = \nabla_X Y + h(X,Y), \quad \bar{\nabla}_X U = -A_U X + \nabla^t_X U,$

for any $X, Y \in \Gamma(TM)$ and $U \in \Gamma(tr(TM))$, where $\{\nabla_X Y, A_U X\}$ and $\{h(X,Y), \nabla^t_X U\}$ belong to $\Gamma(TM)$ and $\Gamma(tr(TM))$, respectively. Here ∇ is a torsion-free linear connection on M, h is a symmetric bilinear form on $\Gamma(TM)$ which is called second fundamental form, A_U is a linear operator on M and known as shape operator.

According to (2.1), considering the projection morphisms L and S of $tr(TM)$ on $ltr(TM)$ and $S(TM^\perp)$, respectively, then (2.3) becomes

(2.4)
$\bar{\nabla}_X Y = \nabla_X Y + h^l(X,Y) + h^s(X,Y), \quad \bar{\nabla}_X U = -A_U X + D^l_X U + D^s_X U,$

where we put $h^l(X,Y) = L(h(X,Y))$, $h^s(X,Y) = S(h(X,Y))$, $D^l_X U = L(\nabla^t_X U)$, $D^s_X U = S(\nabla^t_X U)$.

As h^l and h^s are $\Gamma(ltr(TM))$-valued and $\Gamma(S(TM^\perp))$-valued respectively, therefore they are called as the lightlike second fundamental form and the screen second fundamental form on M. In particular

(2.5) $\quad\quad \bar{\nabla}_X N = -A_N X + \nabla^l_X N + D^s(X,N),$

(2.6) $\quad\quad \bar{\nabla}_X W = -A_W X + \nabla^s_X W + D^l(X,W),$

where $X \in \Gamma(TM), N \in \Gamma(ltr(TM))$ and $W \in \Gamma(S(TM^\perp))$. Using (2.4)-(2.6) we obtain

$$\bar{g}(h^s(X,Y),W) + \bar{g}(Y,D^l(X,W)) = g(A_W X, Y),$$

$$\bar{g}(h^l(X,Y),\xi) + \bar{g}(Y,h^l(X,\xi)) + g(Y,\nabla_X \xi) = 0,$$

for any $\xi \in \Gamma(RadTM), W \in \Gamma(S(TM^\perp))$ and $N, N' \in \Gamma(ltr(TM))$.

Let P be the projection morphism of TM on $S(TM)$ then we can induce some new geometric objects on the screen distribution $S(TM)$ on M as

(2.7) $$\nabla_X PY = \nabla_X^* PY + h^*(X,Y),$$

(2.8) $$\nabla_X \xi = -A_\xi^* X + \nabla_X^{*t} \xi,$$

for any $X, Y \in \Gamma(TM)$ and $\xi \in \Gamma(RadTM)$, where $\{\nabla_X^* PY, A_\xi^* X\}$ and $\{h^*(X,Y), \nabla_X^{*t} \xi\}$ belong to $\Gamma(S(TM))$ and $\Gamma(RadTM)$, respectively. ∇^* and ∇^{*t} are linear connections on complementary distributions $S(TM)$ and $RadTM$, respectively. h^* and A^* are $\Gamma(RadTM)$-valued and $\Gamma(S(TM))$-valued bilinear forms and called as the second fundamental forms of distributions $S(TM)$ and $RadTM$, respectively. Using (2.4), (2.7) and (2.8), we obtain

$$\bar{g}(h^l(X,PY),\xi) = g(A_\xi^* X, PY), \quad \bar{g}(h^*(X,PY),N) = \bar{g}(A_N X, PY),$$

for any $X, Y \in \Gamma(TM), \xi \in \Gamma(Rad(TM))$ and $N \in \Gamma(ltr(TM))$.

From the geometry of Riemannian submanifolds and non degenerate submanifolds, it is known that the induced connection ∇ on a non degenerate submanifold is a metric connection. Unfortunately, this is not true for a lightlike submanifold. Indeed, considering $\bar{\nabla}$ a metric connection then we have

$$(\nabla_X g)(Y,Z) = \bar{g}(h^l(X,Y),Z) + \bar{g}(h^l(X,Z),Y),$$

for any $X, Y, Z \in \Gamma(TM)$.

In their celebrated paper [1], Barros and Romero defined indefinite Kaehler manifolds as

Definition 2.2. *Let $(\bar{M}, J, \bar{g})$ be an indefinite almost Hermitian manifold and $\bar{\nabla}$ be the Levi-Civita connection on $\bar{M}$ with respect to $\bar{g}$. Then $\bar{M}$ is called an indefinite Kaehler manifold if J is parallel with respect to $\bar{\nabla}$, that is*

(2.9) $$(\bar{\nabla}_X J)Y = 0, \quad \forall \ X, Y \in \Gamma(T\bar{M}).$$

Indefinite complex space form is a connected indefinite Kaehler manifold of constant holomorphic sectional curvature c, denoted by $\bar{M}(c)$, whose curvature tensor field $\bar{R}$ is given by [1].

$$\bar{R}(X,Y)Z = \frac{c}{4}\{\bar{g}(Y,Z)X - \bar{g}(X,Z)Y + \bar{g}(JY,Z)JX$$
(2.10)
$$-\bar{g}(JX,Z)JY + 2\bar{g}(X,JY)JZ\}.$$

for any $X, Y \in \Gamma(T\bar{M})$

3. Generic Lightlike Submanifolds

There exists a class of submanifolds, called generic submanifolds, of an almost complex manifold $\bar{M}$ with induced non-degenerate metric g. We say that M is a generic submanifold of $\bar{M}$ if the normal bundle $(TM)^\perp$ of M is mapped into the tangent bundle TM by action of the structure tensor J of $\bar{M}$, that is, $J(TM)^\perp \subset TM$ [6, 7]. In this section we extend the concept of generic submanifold to an almost complex manifold $\bar{M}$ having degenerate (lightlike) induced metric g.

Kupeli [5] obtained that the screen distribution $S(TM)$ is not unique since it is canonically isomorphic to the factor vector bundle $S(TM)^* = TM/Rad(TM)$. Therefore all screen distributions $S(TM)$ are mutually isomorphic. Moreover all the screen distributions are non-degenerate. Hence, we define generic lightlike submanifolds of an indefinite almost complex manifold $\bar{M}$ as follow:

Definition 3.1. *We say that M is a generic lightlike submanifold of an indefinite almost complex manifold $\bar{M}$ if there exists a screen distribution $S(TM)$ of M such that*

(3.1)
$$J(S(TM)^\perp) \subset S(TM).$$

Let $(\bar{M}, \bar{g}, J)$ be a real $2m$-dimensional, $m > 1$, indefinite almost Hermitian manifold, where $\bar{g}$ is a semi- Riemannian metric of index $\nu = 2q, 0 < q < m$. Let (M, g) be a lightlike subamnifold of $\bar{M}$, where g is the degenerate induced metric on M. Since the ambient manifold $\bar{M}$ has additional geometric structure J, we expect a particular screen distribution $S(TM)$ on M. Let ξ be a local section of $Rad(TM)$ then $\bar{g}(J\xi, \xi) = 0$ implies that $J\xi$ is tangent to M, that is, $J(Rad(TM))$ is a distribution on M such that $Rad(TM) \cap J(Rad(TM)) = \{0\}$. Let N be a local section of the lightlike transversal vector bundle $ltr(TM)$ of M then

$\bar{g}(JN,\xi) = -\bar{g}(N,J\xi) = 0$, this implies that JN is tangent to M. Moreover, $\bar{g}(JN,N) = 0$ implies that the component of JN with respect to ξ vanishes and therefore $JN \in \Gamma(S(TM))$. Since $\bar{g}(JN,J\xi) = \bar{g}(N,\xi) = 1$ therefore $J(ltr(TM)) \oplus J(Rad(TM))$ is a vector subbundle of $S(TM)$. Then there exists a non degenerate distribution μ on M such that

(3.2)
$$TM = \{Jltr(TM) \oplus JRad(TM)\} \perp JS(TM^\perp) \perp Rad(TM) \perp \mu,$$

further we can write (3.2) as

(3.3) $$TM = D \oplus D'$$

where $D = Rad(TM) \perp JRad(TM) \perp \mu$ and $D' = \{Jltr(TM) \perp JS(TM^\perp)\}$. Let P, Q_1 and Q_2 be the projections from TM to D, $Jltr(TM)$ and $JS(TM^\perp)$ respectively then any X in TM can be written as

(3.4) $$X = PX + Q_1X + Q_2X,$$

applying J to (3.4) we obtain

(3.5) $$JX = TX + w_1X + w_2X,$$

where $TX = JPX \in D$, $w_1X = JQ_1X \in ltr(TM)$ and $w_2X = JQ_2X \in S(TM^\perp)$. Then (3.5) can be written as

(3.6) $$JX = TX + wX,$$

where TX and wX are the tangential and transversal components of JX, respectively.
Similarly

(3.7) $$JV = BV,$$

for any $V \in \Gamma(tr(TM))$, where BV is the sections of TM

Differentiating (3.5) and using (2.4)-(2.6) and (3.7) we have

(3.8) $$D^s(X, w_1Y) = -\nabla^s_X w_2Y + w_2\nabla_X Y - h^s(X, TY).$$

(3.9) $$D^l(X, w_2Y) = -\nabla^l_X w_1Y + w_1\nabla_X Y - h^l(X, TY).$$

Using Kaehlerian property of $\bar{\nabla}$ with (2.5) and (2.6), we have the following lemmas.

Lemma 3.2. *Let M be a Genric lightlike submanifold of an indefinite Kaehlerian manifold $\bar{M}$. Then we have*

(3.10) $\quad (\nabla_X T)Y = A_{w_1 Y} X + A_{w_2 Y} X + Bh^s(X,Y) + Bh^l(X,Y),$

(3.11) $\quad (\nabla^l_X w_1)Y = -h^l(X, TY) - D^l(X, w_2 Y),$

and

(3.12) $\quad (\nabla^s_X w_2)Y = -h^s(X, TY) - D^s(X, w_1 Y).$

where $X, Y \in \Gamma(TM)$ and

(3.13) $\quad (\nabla_X T)Y = \nabla_X TY - T\nabla_X Y,$

(3.14) $\quad (\nabla^l_X w_1)Y = \nabla^l_X w_1 Y - w_1 \nabla_X Y,$

and

(3.15) $\quad (\nabla^s_X w_2)Y = \nabla^s_X w_2 Y - w_2 \nabla_X Y.$

Lemma 3.3. *Let M be a Genric lightlike submanifold of an indefinite Kaehlerian manifold $\bar{M}$. Then we have*

(3.16) $\quad (\nabla^s_X B)W = -TA_W X + BD^l(X, W),$

and

(3.17) $\quad (\nabla^l_X B)N = -TA_N X + BD^s(X, N),$

where $X \in \Gamma(TM)$, $W \in \Gamma(S(TM^\perp))$, $N \in \Gamma(ltr(TM))$ and

(3.18) $\quad (\nabla^s_X B)W = \nabla_X BW - B\nabla^s_X W,$

(3.19) $\quad (\nabla^l_X B)N = \nabla_X BN - B\nabla^l_X N.$

4. Existence Theorem

Theorem 4.1. *A lightlike submanifold M of an indefinite complex space form $\bar{M}(c)$ with $c \neq 0$ is genric lightlike submanifold with $\mu \neq 0$ if and only if*

(a) *The maximal subspace of $T_p M, p \in M$ define a distribution*

$$D = Rad(TM) \perp JRad(TM) \perp \mu,$$

where μ is a non-degenerate complex distribution.

(b) *There exists a lightlike transversal vector bundle ltr(TM) such that*
$$\bar{g}(\bar{R}(X,Y)N, N') = 0,$$
for any $X, Y \in \Gamma(\mu)$ *and* $N, N' \in \Gamma(ltr(TM))$.

(c) *There exists a screen transversal lightlike vector bundle* $S(TM^\perp)$ *such that*
$$\bar{g}(\bar{R}(X,Y)W, W') = 0,$$
for any $X, Y \in \Gamma(\mu)$ *and* $W, W' \in \Gamma(S(TM^\perp))$.

Proof. Suppose M be a genric lightlike submanifold of $\bar{M}(c)$ such that $c \neq 0$. Then, using the definition of generic lightlike submanifold
$$D = Rad(TM) \perp JRad(TM) \perp \mu,$$
is a maximal invariant subspace. Let $X, Y \in \Gamma(\mu)$ and $N, N' \in \Gamma(ltr(TM))$ then using (3.18), we have
$$g(\bar{R}(X,Y)N, N') = \frac{c}{2}g(X, JY)g(JN, N'),$$
using the definition of genric lightlike submanifolds, we have $\bar{g}(\bar{J}N, N') = 0$. Therefore $\bar{g}(\bar{R}(X,Y)N, N') = 0$. Similarly for any $X, Y \in \Gamma(\mu)$ and $W, W' \in \Gamma(S(TM^\perp))$, using (3.18), we get
$$\bar{g}(\bar{R}(X,Y)W, W') = \frac{c}{2}\bar{g}(X, JY)\bar{g}(W', W) = 0.$$
Conversely, assume that (a), (b) and (c) are satisfied. Then from (a), it is clear that $Rad(TM)$ is a distribution of TM such that $Rad(TM) \cap JRad(TM) = \{0\}$. This implies that $JRad(TM)$ is tangent to M. Now, using (b) and (3.18), we have
$$\frac{c}{2}\bar{g}(X, JY)\bar{g}(JN, N') = 0,$$
for any $X, Y \in \Gamma(\mu)$ and $N, N' \in \Gamma(ltr(TM))$. Since $c \neq \{0\}$ and μ is non-degenerate therefore we have

(4.1) $$\bar{g}(JN, N') = 0,$$

that is, component of JN with respect to ξ vanishes and $J(ltr(TM))$ defines a distribution on M such that $J(ltr(TM)) \cap Rad(TM) = \{0\}$. On the other hand, using (c) and (3.18), we have
$$\frac{c}{2}\bar{g}(X, JY)\bar{g}(JW, W') = 0,$$

for any $X, Y \in \Gamma(\mu)$ and $W, W' \in \Gamma(S(TM^\perp))$. Again since $c \neq \{0\}$ and μ is non-degenerate therefore we get

(4.2) $$\bar{g}(JW, W') = 0,$$

that is, $J(S(TM^\perp)) \perp S(TM^\perp)$. Similarly, we can prove that $J(S(TM^\perp)) \perp ltr(TM)$, $J(S(TM^\perp)) \perp Rad(TM)$, $J(S(TM^\perp)) \perp J(Rad(TM))$, $J(S(TM^\perp)) \perp \mu$ and $J(S(TM^\perp)) \perp J(ltr(TM))$. Thus we can conclude that $J(S(TM^\perp)) \subset S(TM)$, this completes the proof. □

Theorem 4.2. *Let M be a genric lightlike submanifold of an indefinite Kaehler manifold $\bar{M}$ then*

(i) *the distribution D is integrable, if and only if, $h(X, TY) = h(Y, TX)$, for any $X, Y \in \Gamma(D)$.*

(ii) *the distribution D' is integrable, if and only if, $A_{JZ}U = A_{JU}Z$, for any $Z, U \in \Gamma(D')$.*

Proof. Let $X, Y \in \Gamma(D)$ then using (3.11), (3.12), (3.14) and (3.15), we have

$$\omega_1 \nabla_X Y = h^l(X, TY), \quad \text{and} \quad \omega_2 \nabla_X Y = h^s(X, TY),$$

that is,

$$\omega \nabla_X Y = h(X, TY).$$

Replacing X by Y and then subtracting the resulting equation from this equation, we get

$$\omega[X, Y] = h(X, TY) - h(Y, TX),$$

which proves (i). Now, let $Z, U \in \Gamma(D')$ then using (3.10) and (3.13), we have

$$-T\nabla_Z U = A_{w_1 U} Z + A_{w_2 U} Z + Bh^s(Z, U) + Bh^l(Z, U) = A_{\omega U} Z + Bh(Z, U),$$

Then, similarly as above, we have

$$T[Z, U] = A_{JZ}U - A_{JU}Z,$$

this completes the proof of (ii). □

Theorem 4.3. *Let M be a genric lightlike submanifold of an indefinite Kaehler manifold $\bar{M}$. Then the distribution D defines a totally geodesic foliation in M, if and only if, $h(X, JY) = 0$, for any $X, Y \in \Gamma(D)$.*

Proof. Since $D' = Jltr(TM) \perp JS(TM^\perp)$, therefore D defines a totally geodesic foliation in M, if and only if

$$g(\nabla_X Y, J\xi) = g(\nabla_X Y, JW) = 0,$$

for any $X, Y \in \Gamma(D)$, $\xi \in \Gamma(Rad(TM))$ and $W \in \Gamma(S(TM^\perp))$. Using (2.4), we have

(4.3) $\qquad g(\nabla_X Y, J\xi) = -\bar{g}(\bar{\nabla}_X JY, \xi) = -\bar{g}(h^l(X, JY), \xi),$

and

(4.4) $\qquad g(\nabla_X Y, JW) = -\bar{g}(\bar{\nabla}_X JY, W) = -\bar{g}(h^s(X, JY), W).$

Hence, from (4.3) and (4.4), the assertion follows. $\square$

Theorem 4.4. *Let M be a genric lightlike submanifold of an indefinite Kaehler manifold $\bar{M}$. Then the distribution D' defines a totally geodesic foliation in M, if and only if, $A_{\omega Y} X \in \Gamma(D')$, for any $X, Y \in \Gamma(D')$.*

Proof. Using (3.10) and (3.13), we have

$$-T\nabla_X Y = A_{\omega Y} X + Bh(X, Y),$$

for any $X, Y \in \Gamma(D')$. Let D' defines a totally geodesic foliation in M, therefore $A_{\omega Y} X = -Bh(X, Y)$ and this implies that $A_{\omega Y} X \in \Gamma(D')$. Conversely, $A_{\omega Y} X \in \Gamma(D')$, for any $X, Y \in \Gamma(D')$ implies that $T\nabla_X Y = 0$. Hence $\nabla_X Y \in \Gamma(D')$ which completes the proof.
$\square$

Theorem 4.5. *Let M be a genric lightlike submanifold of an indefinite Keahler manifold $\bar{M}$. Then the induced connection ∇ is metric connection, if and only if,*

$$\nabla^*_X J\xi \in \Gamma(JRad(TM)), \quad h^*(X, J\xi) = 0 \quad \text{and} \quad Bh(X, J\xi) = 0,$$

for any $\xi \in \Gamma(Rad(TM))$ and $X \in \Gamma(TM)$.

Proof. Let $X \in \Gamma(TM)$ and $\xi \in \Gamma(Rad(TM))$ then using (2.9), we have

$$\bar{\nabla}_X J\xi = J\bar{\nabla}_X \xi,$$

then using (2.3), we obtain

$$\nabla_X \xi + h(X, \xi) = -J(\nabla_X J\xi + h(X, J\xi)).$$

Since $\xi \in \Gamma(Rad(TM))$ therefore $J\xi \in \Gamma(S(TM))$, hence using the definition of genric lightlike submanifolds, we obtain

$$\nabla_X \xi + h(X,\xi) = -T\nabla_X J\xi - \omega_1 \nabla_X J\xi - \omega_2 \nabla_X J\xi - Bh(X, J\xi).$$

Equating tangential components of above equation both sides, we get

$$\nabla_X \xi = -T\nabla_X J\xi - Bh(X, J\xi),$$

now using (2.7), we obtain

$$\nabla_X \xi = -T\nabla_X^{*t} J\xi - Th^*(X, JY) - Bh(X, J\xi).$$

Thus from above equation, it is clear that, $\nabla_X \xi \in \Gamma(RadTM)$, if and only if, $Bh(X, J\xi) = 0$, $\nabla_X^{*t} J\xi \in \Gamma(J(RadTM))$ and $h^*(X, JY) = 0$. Hence, the assertion follows using the Theorem 2.4 in [2], page no. 161. □

REFERENCES

[1] Barros, M. and Romero, A., *Indefinite Kaehler manifolds*, Math. Ann., **261** (1982), 55–62.

[2] Duggal, K. L. and Bejancu, A., *Lightlike Submanifolds of semi-Riemannian Manifolds and Applications*, Vol. 364 of Mathematics and its Applications, Kluwer Academic Publishers, Dordrecht, The Netherlands, (1996).

[3] Duggal, K. L. and Jin, D.H., *Generic lightlike submanifolds of an indefinite Sasakian manifold*, Int. Electron. J. Geom., **5**(1), (2012), 108-119.

[4] Jin, D.H. and Lee, J.W., *Generic lightlike submanifolds of an indefinite cosymplectic manifold*, Math. Probl. Eng., Volume 2011, Article ID 610986, 16 pages, 2011.

[5] Kupeli, D. N., *Singular semi-Riemannian Geometry*, Kluwer, Dordrecht, (1996).

[6] Yano, K. and Kon, M., *Generic submanifolds*, Ann. di Math. pura Appl., **123**(1980), 59-92.

[7] Yano, K. and Kon, M., *Generic submanifolds of Sasakian manifolds*, Kodai Math. J., **3**(1980), 163-196.

Manish Gogna
Baba Banda Singh Bahadur Engineering College, Fathehgarh Sahib, India.
E-mail: manish_bbsbec@yahoo.co.in

Sangeet Kumar
Department of Applied Sciences,

Chitkara University, Jhansla, Distt. Rajpura, Patiala, India.
E-mail: sp7maths@gmail.com

Rakesh Kumar[1]
University College of Engineering, Punjabi University, Patiala, India.
E-mail: dr_rk37c@yahoo.co.in
&
R. K. Nagaich
Department of Mathematics, Punjabi University, Patiala, India.
E-mail: rakeshnagaich@yahoo.com

---

[1]Corresponding Author

ON THE W_2-CURVATURE TENSOR OF $N(k)$-CONTACT METRIC MANIFOLDS

SHYAMAL KUMAR HUI

ABSTRACT. The object of the present paper is to study $N(k)$-contact metric manifolds admitting W_2-curvature tensor. We classify $N(k)$-contact metric manifolds satisfying the conditions $R(\xi, U) \cdot W_2 = 0$, $W_2(\xi, U) \cdot R = 0$ and $W_2(\xi, X) \cdot S = 0$. In this paper, ξ-W_2 flat and ϕ-W_2 flat $N(k)$-contact metric manifolds are also studied.

1. INTRODUCTION

A contact manifold is a smooth $(2n+1)$-dimensional manifold M^{2n+1} equipped with a global 1-form η such that $\eta \wedge (d\eta)^n \neq 0$ everywhere. Given a contact form η, there exists a unique vector field ξ, called the characteristic vector field of η, satisfying $\eta(\xi) = 1$ and $d\eta(X, \xi) = 0$ for any vector field X on M^{2n+1}. A Riemannian metric g is said to be associated metric if there exists a tensor field ϕ of type $(1,1)$ such that

(1.1) $\quad \eta(X) = g(X, \xi), d\eta(X, Y) = g(X, \phi Y) = -g(\phi X, Y)$

$$\phi^2 X = -X + \eta(X)\xi$$

for all vector fields X, Y on M^{2n+1}. Then the structure (ϕ, ξ, η, g) on M^{2n+1} is called a contact metric structure and the manifold M^{2n+1} equipped with such a structure is said to be a contact metric manifold [3]. It can be easily seen that in a contact metric manifold, the following relations hold:

(1.2) $\quad \phi\xi = 0, \quad \eta(\phi X) = 0, \quad g(\phi X, \phi Y) = g(X, Y) - \eta(X)\eta(Y)$

for any vector field X, Y on M^{2n+1}.

2000 *Mathematics Subject Classification.* 53C15, 53C25.

Key words and phrases. Contact metric manifold, $N(k)$-contact metric manifold, W_2-curvature tensor, Einstein manifold, η-Einstein manifold.

In a contact metric manifold $M^{2n+1}(\phi, \xi, \eta, g)$, we define a (1,1) tensor field h by $h = \frac{1}{2}\pounds_\xi \phi$, where $\pounds$ denotes the operator of Lie differentiation. Then h is symmetric and satisfies

(1.3) $\qquad h\xi = 0, \quad h\phi = -\phi h, \quad Tr.h = Tr.\phi h = 0.$

Also we have the following relation

(1.4) $\qquad\qquad \nabla_X \xi = -\phi X - \phi h X,$

where ∇ denotes the Riemannian connection of g.

A contact metric manifold $M^{2n+1}(\phi, \xi, \eta, g)$ for which ξ is a Killing vector field is called a K-contact manifold.

In 1988 Tanno [26] introduced the notion of k-nullity distribution of a contact metric manifold as a distribution such that the characteristic vector field ξ of the contact metric manifold belongs to the distribution. The contact metric manifold with ξ belonging to the k-nullity distribution is called $N(k)$-contact metric manifold. The $N(k)$-contact metric manifold is also studied by De and Gaji [8], De and Mondal [9], Ghosh, De and Taleshian [12], Shaikh and Bagewadi [21], Yildiz et. al [30] and many others.

In 1970 Pokhariyal and Mishra [20] introduced new tensor fields, called W_2 and E tensor fields, in a Riemannian manifold and studied their properties. According to them a W_2-curvature tensor on a manifold $(M^{2n+1}, g), n > 1$, is defined by [20]

(1.5) $\quad W_2(X,Y)Z = R(X,Y)Z + \frac{1}{2n}\big[g(X,Z)QY - g(Y,Z)QX\big],$

where Q is the Ricci-operator, i.e., $g(QX, Y) = S(X, Y)$ for all X, Y.

The W_2-curvature tensor was introduced on the line of Weyl projective curvature tensor and by breaking W_2 into skew-symmetric parts the tensor E has been defined. Rainich conditions for the existence of the non-null electrovariance can be obtained by W_2 and E, if we replace the matter tensor by the contracted part of these tensors. The tensor E enables to extend Pirani formulation of gravitational waves to Einstein space ([18], [19]). It is shown that [20] except the vanishing of complexion vector and property of being identical in two spaces which are in geodesic correspondence, the W_2-curvature tensor possesses the properties almost similar to the Weyl projective curvature tensor. Thus we can very well use

W_2-curvature tensor in various physical and geometrical spheres in place of the Weyl projective curvature tensor.

The W_2-curvature tensor have also been studied by various authors in different structures such as De and Sarkar [10], Hui and Sarkar [13], Matsumoto, Ianus and Mihai [15], Pokhariyal ([17], [18], [19]), Shaikh, Jana and Eyasmin [22], Shaikh, Matsuyama and Jana [23], Taleshian and Hosseinzadeh [25], Tripathi and Gupta [27], Venkatesha, Bagewadi and Kumar [28], Yildiz and De [29], Yildiz et. al [30] and many others.

In [5] Blair, Kim and Tripathi studied the concircular curvature tensor of a contact metric manifold. In [16] Pak and Shin studied conformal curvature tensor of a contact metric manifold. Also Kim et. al [14] studied conformal curvature tensor of a contact metric manifold. In [30] Yildiz et. al studied the Weyl projective curvature tensor of an $N(k)$-contact metric manifolds. Recntly Ghosh, De and Taleshian [12] studied conharmonic curvature tensor on an $N(k)$-contact metric manifolds. Motivated by the above studies, the object of the present paper is to study W_2-curvature tensor of $N(k)$-contact metric manifolds. The paper is organized as follows. Section 2 is concerned with $N(k)$-contact metric manifolds.

A Riemannian manifold (M^{2n+1}, g) is said to be semisymmetric [24] if it satisfies $R(X,Y) \cdot R = 0$, where $R(X,Y)$ acts as a derivation on R. Section 3 is devoted to the study of $N(k)$-contact metric manifolds with $R(\xi, U) \cdot W_2 = 0$. It is proved that if a $N(k)$-contact metric manifold satisfies $R(\xi, U) \cdot W_2 = 0$, then the manifold is either locally isometric to the Riemannian product $E^{n+1}(0) \times S^n(4)$ or Einstein. Section 4 deals with $N(k)$-contact metric manifolds with $W_2(\xi, U) \cdot R = 0$.

A Riemannian manifold (M^{2n+1}, g) is said to be Ricci-semisymmetric [24] if its Ricci tensor S satisfies $R(X,Y) \cdot S = 0$, where $R(X,Y)$ acts as a derivation on S. Ricci-semisymmetric manifold are studied by several authors. Section 5 consists with a study of $N(k)$-contact metric manifolds satisfying $W_2(\xi, X) \cdot S = 0$. It is shown that if a $N(k)$-contact metric manifold satisies $W_2(\xi, X) \cdot S = 0$ then the square of the length of the Ricci tensor of such a manifold is $2nkr$, where r is the scalar curvature of the manifold.

A Riemannian manifold (M^{2n+1}, g) is said to be flat if $R(X,Y)Z = 0$. It is called ξ-flat if $R(X,Y)\xi = 0$, where ξ is a non-null unit vector field in M^{2n+1}. The condition of ξ-flatness is weaker than the condition of flatness. In [7], De and Biswas studied the ξ-conformally flat contact metric manifolds with $\xi \in N(k)$ is ξ-conformally flat if and only if it is an η-Einstein manifold. In [11] Dwivedi and Kim proved that a Sasakian manifold is ξ-conharmonically flat if and only if it is an η-Einstein. Section 6 deals with the study of ξ-W_2 flat $N(k)$-contact metric manifolds. It is shown that a $N(k)$-contact metric manifold is ξ-W_2 flat if and only if it is an Einstein manifold.

In [12] Ghosh, De and Taleshian studied ϕ-conharmonically flat $N(k)$-contact metric manifolds. In section 7, we have studied ϕ-W_2 flat $N(k)$-contact metric manifolds. A $(2n+1)$-dimensional $N(k)$-contact metric manifold is said to be ϕ-W_2 flat if $W_2(\phi X, \phi Y, \phi Z, \phi W) = 0$ for any vector field X, Y, Z and $W \in M$. It is shown that a $(2n+1)$-dimensional ϕ-W_2 flat $N(k)$-contact metric manifold is a Sasakian manifold.

2. $N(k)$-CONTACT METRIC MANIFOLDS

Let us consider a contact metric metric manifold $M^{2n+1}(\phi, \xi, \eta, g)$. The k-nullity distribution [26] of a Riemannian manifold (M, g) for

$$N(k) : p \to N_p(k) = \{Z \in T_pM : R(X,Y)Z = k[g(Y,Z)X - g(X,Z)Y]\}$$

for any X, $Y \in T_pM$. Hence if the characteristic vector field ξ of a contact metric manifold belongs to the k-nullity distribution, then we have

(2.1) $$R(X,Y)\xi = k[\eta(Y)X - \eta(X)Y].$$

Thus a contact metric manifold $M^{2n+1}(\phi, \xi, \eta, g)$ satisfying the relation (2.1) is called a $N(k)$-contact metric manifold. In a $N(k)$-contact metric manifold, k is always a constant such that $k \leq 1$ [26]. Also in a $N(k)$-contact metric manifold $M^{2n+1}(\phi, \xi, \eta, g)$, we have the following ([21], [26]):

(2.2) $$Q\phi - \phi Q = 4(n-1)h\phi,$$

(2.3) $$h^2 = (k-1)\phi^2, \quad k \leq 1,$$

(2.4) $$Tr.h^2 = 2n(1-k),$$

(2.5) $$R(\xi, X)Y = k[g(X,Y)\xi - \eta(Y)X],$$

(2.6) $$S(X, \phi Y) + S(\phi X, Y) = 2(2n-2)g(\phi X, hY),$$

(2.7) $S(\phi X, \phi Y) = S(X,Y) - 2nk\eta(X)\eta(Y) - 2(2n-2)g(hX,Y),$

(2.8) $\eta(R(X,Y)Z) = k\{g(Y,Z)\eta(X) - g(X,Z)\eta(Y)\},$

(2.9) $\begin{aligned} S(X,Y) &= 2(n-1)g(X,Y) + 2(n-1)g(hX,Y) \\ &+ [2nk - 2(n-1)]\eta(X)\eta(Y), n \geq 1, \end{aligned}$

(2.10) $S(X,\xi) = 2nk\eta(X), \quad Q\xi = 2nk\xi,$

(2.11) $(\nabla_X \eta)(Y) = g(X + hX, \phi Y),$

(2.12) $\begin{aligned} (\nabla_X h)(Y) &= [(1-k)g(X,\phi Y) + g(X, h\phi Y)]\xi \\ &+ \eta(Y)h(\phi X + \phi h X), \end{aligned}$

(2.13) $(\nabla_X \phi)(Y) = g(X + hX, Y)\xi - \eta(Y)(X + hX)$

for any vector field X, Y on M^{2n+1}. Also in a $N(k)$-contact metric manifold the scalar curvature r is given by ([1], [6], [21])

(2.14) $r = 2n(2n - 2 + k).$

In view of (2.1), (2.4), (2.5) and (2.9), we have from (1.5) that
(2.15)
$$W_2(X,Y)\xi = k[\eta(Y)X - \eta(X)Y] + \frac{1}{2n}[\eta(X)QY - \eta(Y)QX],$$

(2.16) $W_2(\xi, Y)Z = \left[\frac{1}{2n}QY - kY\right]\eta(Z),$

(2.17) $\eta(W_2(X,Y)Z) = 0.$

We recall the following result which will be used later on

Lemma 2.1. [4] *Let $M^{2n+1}(\phi, \xi, \eta, g)$ be a contact metric manifold with $R(X,Y)\xi = 0$ for all vector fields X, Y. Then the manifold is locally isometric to the Riemannian product $E^{n+1}(0) \times S^n(4)$.*

Lemma 2.2. [2] *Let M^{2n+1} be an η-Einstein manifold of dimension $(2n+1)(n \geq 1)$. If ξ belongs to the k-nullity distribution, then $k = 1$ and the structure is Sasakian.*

In a $(2n+1)$-dimensional almost contact metric manifold if $\{e_1, e_2, \cdots, e_{2n}, \xi\}$ is a local orthonormal basis of the tangent space of the manifold then $\{\phi e_1, \phi e_2, \cdots, \phi e_{2n}, \xi\}$ is also a local orthonormal basis. Also we get [12]

$$(2.18) \quad \sum_{i=1}^{2n} g(e_i, e_i) = \sum_{i=1}^{2n} g(\phi e_i, \phi e_i) = 2n,$$

$$(2.19) \quad \sum_{i=1}^{2n} S(e_i, e_i) = \sum_{i=1}^{2n} S(\phi e_i, \phi e_i) = r - 2nk,$$

$$(2.20) \sum_{i=1}^{2n} g(e_i, Z)S(Y, e_i) = \sum_{i=1}^{2n} g(\phi e_i, Z)S(Y, \phi e_i) = S(Y, Z) - 2nk\eta(Y)\eta(Z),$$

$$(2.21) \sum_{i=1}^{2n} g(e_i, \phi Z)S(Y, e_i) = \sum_{i=1}^{2n} g(\phi e_i, \phi Z)S(Y, \phi e_i) = S(Y, \phi Z).$$

Definition 2.1. *A $(2n+1)$-dimensional $N(k)$-contact metric manifold $M^{2n+1}(\phi, \xi, \eta, g)$ $(n > 1)$ is said to be η-Einstein if its Ricci tensor S of type (0,2) is of the form*

$$(2.22) \quad S = ag + b\eta \otimes \eta,$$

where a and b are smooth functions on M.

3. $N(k)$-CONTACT METRIC MANIFOLDS SATISFYING $R(\xi, U) \cdot W_2 = 0$

Let us take a $N(k)$-contact metric manifold $M^{2n+1}(\phi, \xi, \eta, g)(n > 1)$ with $R(\xi, U) \cdot W_2 = 0$, which implies

$$(3.1) \quad R(\xi, U)W_2(X, Y)Z - W_2(R(\xi, U)X, Y)Z \\ - W_2(X, R(\xi, U)Y)Z - W_2(X, Y)R(\xi, U)Z = 0.$$

By virtue of (2.5) we have from (3.1) that

$$(3.2)\ k\big[g(U, W_2(X,Y)Z)\xi - \eta(W_2(X,Y)Z)U - g(U,X)W_2(\xi,Y)Z \\ + \eta(X)W_2(U,Y)Z - g(U,Y)W_2(X,\xi)Z + \eta(Y)W_2(X,U)Z \\ - g(U,Z)W_2(X,Y)\xi + \eta(Z)W_2(X,Y)U\big] = 0.$$

Setting $Z = \xi$ in (3.2) and using (2.15) and (2.16), we get

$$k\big[W_2(X,Y)U - g(U,Y)\{kX - \frac{1}{2n}QX\}$$
$$-k\{\eta(X)g(Y,U) - \eta(Y)g(X,U)\}\xi$$
$$-\frac{1}{2n}\{\eta(Y)S(X,U) - \eta(X)S(Y,U)\}\xi\big] = 0,$$

which implies either $k = 0$ or

$$(3.3) \quad W_2(X,Y)U = g(U,Y)\{kX - \frac{1}{2n}QX\}$$
$$+ k\{\eta(X)g(Y,U) - \eta(Y)g(X,U)\}\xi$$
$$+ \frac{1}{2n}\{\eta(Y)S(X,U) - \eta(X)S(Y,U)\}\xi.$$

If $k = 0$ then from (2.1), we have $R(X,Y)\xi = 0$ for all X, Y and hence by Lemma 2.1, it follows that the manifold is locally isometric to the Riemannian product $E^{n+1}(0) \times S^n(4)$.

Next we consider the case (3.3). Setting $U = \xi$ in (3.3) and using (1.1) and (2.10), we get

$$(3.4) \quad W_2(X,Y)\xi = \eta(Y)\big[kX - \frac{1}{2n}QX\big].$$

In view of (2.15), (3.4) yields

$$QY = 2nkY,$$

i.e.,

$$(3.5) \quad S(Y,V) = 2nkg(Y,V),$$

which implies that the manifold under consideration is Einstein. Thus we can state the following:

Theorem 3.1. *Let $M^{2n+1}(\phi, \xi, \eta, g)(n > 1)$ be a $(2n+1)$-dimensional $N(k)$-contact metric manifolds with $R(\xi, U) \cdot W_2 = 0$. Then the manifold is either locally isometric to the Riemannian product $E^{n+1}(0) \times S^n(4)$ or is Einstein.*

4. $N(k)$-CONTACT METRIC MANIFOLDS WITH $W_2(\xi, U) \cdot R = 0$

We now consider a $N(k)$-contact metric manifold $M^{2n+1}(\phi, \xi, \eta, g)(n > 1)$ satisfying $W_2(\xi, U) \cdot R = 0$, which implies

(4.1) $$W_2(\xi, U)R(X,Y)Z - R(W_2(\xi, U)X, Y)Z \\ - R(X, W_2(\xi, U)Y)Z - R(X,Y)W_2(\xi, U)Z = 0.$$

In view of (2.8) and (2.16), (4.1) yields

$$k[\eta(X)g(Y,Z) - \eta(Y)g(X,Z)][\frac{1}{2n}QU - kU] \\ + k[\eta(X)R(U,Y)Z + \eta(Y)R(X,U)Z + \eta(Z)R(X,Y)U] \\ - \frac{1}{2n}[\eta(X)R(QU,Y)Z + \eta(Y)R(X,QU)Z + \eta(Z)R(X,Y)QU] = 0.$$

Setting $Z = \xi$ in (4.2) and using (2.1), we get

(4.3) $$R(X,Y)QU = 2nkR(X,Y)U,$$

i.e.,

(4.4) $$R(X,Y,QU,V) = 2nkR(X,Y,U,V).$$

Contracting (4.4) over X and V, we get

(4.5) $$S(Y, QU) = 2nkS(Y, U).$$

Let $\{e_1, e_2, \cdots, e_{2n}, \xi\}$ be a local orthonormal basis of the tangent space of the manifold and let $l^2 = \sum_{i=1}^{2n+1} S(e_i, Qe_i)$ be the square of the length of Ricci tensor. Then from (4.5), we get

(4.6) $$l^2 = \sum_{i=1}^{2n+1} S(e_i, Qe_i) = 2nk \sum_{i=1}^{2n+1} S(e_i, e_i) = 2nkr,$$

where r is the scalar curvature of the manifold.
This leads to the following:

Theorem 4.1. *The square of the length of the Ricci tensor of a $(2n+1)$-dimensional $N(k)$-contact metric manifold $M^{2n+1}(\phi, \xi, \eta, g)$ $(n > 1)$ with $W_2(\xi, U) \cdot R = 0$ is $2nkr$.*

5. $N(k)$-CONTACT METRIC MANIFOLDS SATISFYING $W_2(\xi, X) \cdot S = 0$

We now consider a $N(k)$-contact metric manifold $M^{2n+1}(\phi, \xi, \eta, g)(n > 1)$ with $W_2(\xi, X) \cdot S = 0$. Then we have

(5.1) $\qquad S(W_2(X,Y)Z, \xi) + S(Z, W_2(X,Y)\xi) = 0.$

Using (2.10), (2.15) and (2.17) in (5.1), we get
(5.2)
$\eta(X)S(QY, Z) - \eta(Y)S(QX, Z) = 2nk[\eta(X)S(Y,Z) - \eta(Y)S(X,Z)].$

Setting $Z = \xi$ in (5.2) and using (2.10), we obtain

(5.3) $\qquad S(QY, Z) = 2nkS(Y, Z).$

Let $\{e_1, e_2, \cdots, e_{2n}, \xi\}$ be a local orthonormal basis of the tangent space of the manifold and let $l^2 = \sum_{i=1}^{2n+1} S(e_i, Qe_i)$ be the square of the length of Ricci tensor. Then from (4.5), we get

(5.4) $\qquad l^2 = \sum_{i=1}^{2n+1} S(e_i, Qe_i) = 2nk \sum_{i=1}^{2n+1} S(e_i, e_i) = 2nkr,$

where r is the scalar curvature of the manifold.
This leads to the following:

Theorem 5.1. *The square of the length of the Ricci tensor of a $(2n+1)$-dimensional $N(k)$-contact metric manifold $M^{2n+1}(\phi, \xi, \eta, g)$ $(n > 1)$ with $W_2(\xi, U) \cdot S = 0$ is $2nkr$.*

6. ξ-W_2 FLAT $N(k)$-CONTACT METRIC MANIFOLDS

In [12], Ghosh, De and Taleshian studied ξ-conharmonically flat $N(k)$-contact metric manifolds. Motivated by the above studies, in this section we consider a $(2n+1)$-dimensional ξ-W_2 flat $N(k)$-contact metric manifolds. Then from (1.5), we obtain

(6.1) $\qquad R(X,Y)\xi = \frac{1}{2n}[\eta(Y)QX - \eta(X)QY].$

Using (2.1) in (6.1), we obtain

(6.2) $\qquad k[\eta(Y)X - \eta(X)Y] = \frac{1}{2n}[\eta(Y)QX - \eta(X)QY].$

Putting $Y = \xi$ in (6.2) and using (2.10), we obtain

(6.3) $\qquad QX = 2nkX,$

i.e.

(6.4) $$S(X,U) = 2nkg(X,U),$$

which implies that the manifold under consideration is an Einstein. Conversely, we assume that a $(2n+1)$-dimensional $N(k)$-contact metric manifold satisfies the relation (6.4). Then it follows from (1.5) that $W_2(X,Y)\xi = 0$, i.e., the manifold under consideration is ξ-W_2 flat.
Thus we can state the following:

Theorem 6.1. *A $(2n+1)$-dimensional $N(k)$-contact metric manifold is ξ-W_2 flat if and only if it is an Einstein manifold.*

7. ϕ-W_2 FLAT $N(k)$-CONTACT METRIC MANIFOLDS

This section deals with a $(2n+1)$-dimensional ϕ-W_2 flat $N(k)$-contact metric manifold. Then we have from (1.5) that

(7.1)
$$R(\phi X, \phi Y, \phi Z, \phi U) = \frac{1}{2n}\left[g(\phi Y, \phi Z)S(\phi X, \phi U) - g(\phi X, \phi Z)S(\phi Y, \phi U)\right].$$

Let $\{e_1, e_2, \cdots, e_{2n}, \xi\}$ be a local orthonormal basis of the tangent space of the manifold. Then $\{\phi e_1, \phi e_2, \cdots, \phi e_{2n}, \xi\}$ is also a local orthonormal basis of the tangent space. Putting $X = U = e_i$ in (7.1) and summing up from 1 to $2n$, we have

(7.2)
$$\sum_{i=1}^{2n} R(\phi e_i, \phi Y, \phi Z, \phi e_i)$$

$$= \frac{1}{2n}\sum_{i=1}^{2n}\left[g(\phi Y, \phi Z)S(\phi e_i, \phi e_i) - g(\phi e_i, \phi Z)S(\phi Y, \phi e_i)\right].$$

Using (2.19) - (2.21) in (7.2) we obtain

(7.3) $$S(\phi Y, \phi Z) = \frac{r - 2nk}{2n+1}g(\phi Y, \phi Z).$$

Replacing Y and Z by ϕY and ϕZ in (7.3) and using (1.1) and (2.10), we get

(7.4) $$S(Y,Z) = \frac{r - 2nk}{2n+1}g(Y,Z) + \frac{4n(n+1)k - r}{2n+1}\eta(Y)\eta(Z),$$

which implies that the manifold under consideration is an η-Einstein manifold.

This leads to the following:

Theorem 7.1. *A $(2n+1)$-dimensional ϕ-W_2 flat $N(k)$-contact metric manifold is an η-Einstein manifold.*

In view of Lemma 2.2 and Theorem 7.1, we can state the following:

Theorem 7.2. *A $(2n+1)$-dimensional ϕ-W_2 flat $N(k)$-contact metric manifold is a Sasakian manifold.*

REFERENCES

[1] Baikoussis, C., Blair, D. E. and Koufogiorgos, T., *A decomposition of the curvature tensor of a contact manifold satisfying $R(X,Y)\xi = k(\eta(Y)X - \eta(X)Y)$*, Mathematics Technical Report, University of Ioanniana, **1992**.

[2] Baikoussis, C. and Koufogiorgos, T., *On a type of contact manifolds*, J. of Geom., **46** (1993), 1–9.

[3] Blair, D. E., *Contact manifolds in Riemannian geometry*, Lecture Notes in Math. **509**, Springer-Verlag, **1976**.

[4] Blair, D. E., *Two remarks on contact metric structure*, Tohoku Math. J., **29** (1977), 319–324.

[5] Blair, D. E., Kim, J. s. and Tripathi, M. M., *On the concircular curvature tensor of a contact metric manifold*, J. Korean Math. Soc., **42** (2005), 883–892.

[6] Blair, D. E., Koufogiorgos, T. and Papantoniou, B. J., *Contact metric manifolds satisfying a nullity condition*, Israel J. Math., **19** (1995), 189–214.

[7] De, U. C. and Biswas, S., *A note on ξ-conformally flat contact manifolds*, Bull. Malaya. Math. Sci. Soc., **29** (2006), 51–57.

[8] De, U. C. and Gaji, A. K., *On ϕ-recurrent $N(k)$-contact metric manifolds*, Math. J. Okayama Univ., **50** (2008), 101–112.

[9] De, U. C. and Mondal, A. K., *Second order parallel tensor on $N(k)$-contact metric manifolds*, Diff. Geom.-Dynamical Systems, **12** (2010), 158–165.

[10] De, U. C. and Sarkar, A., *On a type of P-Sasakian manifolds*, Math. Reports, **11(61)** (2009), 139–144.

[11] Dwivedi, M. K. and Kim, J.-S., *On conharmonic curvature tensor in K-contact and Sasakian manifolds*, Bull. Malaya. Math. Sci. Soc., **34** (2011), 171–180.

[12] Ghosh, S., De, U. C. and Taleshian, A., *Conharmonic curvature tensor on $N(k)$-contact metric manifolds*, ISRN Geom., **Vol. 2011**, Article ID **423798**.

[13] Hui, S. K. and Sarkar, A., *On the W_2-curvature tensor of generalized Sasakian-space-forms*, Math. Pannonica, **23** (2012), 113–124.

[14] Kim, J. S., Choi, J., Özgür, C. and Tripathi, M. M., *On the contact conformal curvature tensor of a contact metric manifold*, Indian J. Pure Appl. Math., **37** (2006), 199–207.

[15] Matsumoto, K., Ianus, S. and Mihai, I., *On P-Sasakian manifolds which admit certain tensor fields*, Publ. Math. Debrecen, **33** (1986), 61–65.

[16] Pak, J. S. and Shin, Y. J., *A note on contact conformal curvature tensor*, Commun. Korean Math. Soc., **13** (1998), 337–343.

[17] Pokhariyal, G. P., *Study of a new curvature tensor in a Sasakian manifold*, Tensor N. S., **36** (1982), 222–225.

[18] Pokhariyal, G. P., *Relative significance of curvature tensors*, Int. J. Math. and Math. Sci., **5** (1982), 133–139.

[19] Pokhariyal, G. P., *Curvature tensors on A-Einstein Sasakian manifolds*, Balkan J. Geom. Appl., **6** (2001), 45–50.

[20] Pokhariyal, G. P. and Mishra, R. S., *The curvature tensors and their relativistic significance*, Yokohama Math. J., **18** (1970), 105–108.

[21] Shaikh, A. A. and Bagewadi, C. S., *On $N(k)$-contact metric manifolds*, Cubo Math. J., **12** (2010), 183–195.

[22] Shaikh, A. A., Jana, S. K. and Eyasmin, S., *On weakly W_2-symmetric manifolds*, Sarajevo J. Math., **3(15)** (2007), 73–91.

[23] Shaikh, A. A., Matsuyama, Y. and Jana, S. K., *On a type of general relativistic spacetime with W_2-curvature tensor*, Indian J. Math., **50** (2008), 53–62.

[24] Szabó, Z. I., *Structure theorems on Riemannian spaces satisfying $R(X,Y) \cdot R = 0$, The local version*, J. Diff. Geom., **17** (1982), 531–582.

[25] Taleshian, A. and Hosseinzadeh, A. A., *On W_2-curvature tensor $N(k)$-quasi Einstein manifolds*, J. Math. and Computer Science, **1(1)** (2010), 28–32.

[26] Tanno, S., *Ricci curvatures of contact Riemannian manifolds*, Tohoku Math. J., **40** (1988), 441–448.

[27] Tripathi, M. M. and Gupta, P., *On τ-curvature tensor in K-contact and Sasakian manifolds*, Int. Elec. J. Geom., **4** (2011), 32–47.

[28] Venkatesha, Bagewadi, C. S. and Kumar, K. T. Pradeep, *Some results on Lorentzian Para-Sasakian manifolds*, International scholarly research network, doi:10.5402/2011/161523.

[29] Yildiz, A. and De, U. C., *On a type of Kenmotsu manifolds*, Diff. Geom.-Dynamical Systems, **12** (2010), 289–298.

[30] Yildiz, A., De, U. C., Murathan, C. and Arslan, K., *On the Weyl projective curvature tensor of an $N(k)$-contact metric manifold*, Math. Pannonica, **21(1)** (2010), 129–142.

Nikhil Banga Sikshan Mahavidyalaya
Bishnupur, Bankura – 722122
West Bengal. India
E-mail: shyamal_hui@yahoo.co.in

Differential Geometry, Functional Analysis and Applications
Editors: Mohammad Hasan Shahid, Sharfuddin Ahmad *et al.*
Copyright © 2015, Narosa Publishing House, New Delhi

SLANT LIGHTLIKE SUBMANIFOLDS OF INDEFINITE ALMOST CONTACT MANIFOLDS

RASHMI, RAKESH KUMAR AND S. S. BHATIA

ABSTRACT. We study slant lightlike submanifolds of indefinite Sasakian manifolds. We obtain a necessary and sufficient condition for a slant lightlike submanifold to be an anti-invariant. We also derive an equivalent condition for a lightlike submanifold to be a slant lightlike submanifold.

1. INTRODUCTION

The study of Riemannian and semi-Riemannian geometries have been active areas of research in differential geometry. In the case of lightlike submanifolds, the geometry is quite different than the counter part of non-degenerate submanifolds as there is a natural existence of null (lightlike) subspaces. In 1996, Duggal-Bejancu presented a book [6] on the lightlike (degenerate) geometry of submanifolds needed to fill an important missing part in the general theory of submanifolds. Chen [4, 5], introduced the notion of slant submanifolds as a generalization of holomorphic and totally real submanifolds for complex geometry and further extended by Lotta [9] for contact geometry. Cabrerizo et. al. [2, 3] studied slant, semi-slant and bi-slant submanifolds in contact geometry. They all studied the geometry of slant submanifolds with positive definite metric. Therefore this geometry may not be applicable to the other branches of mathematics and physics, where the metric is not necessarily definite. Thus the notion of slant lightlike submanifolds of indefinite Hermitian manifolds was introduced by Sahin [12]. The notion of slant lightlike submanifolds of indefinite Sasakian manifolds is introduced by Sahin and Yildirim in [13], recently and obtained necessary and sufficient conditions for their existence.

2000 *Mathematics Subject Classification.* 53C15, 53C50.
Key words and phrases. Slant lightlike submanifolds, indefinite Sasakian manifolds.

In the present paper, we obtain a necessary and sufficient condition for a slant lightlike submanifold to be an anti-invariant (Theorem (4.1)). We also derive an equivalent condition for a lightlike submanifold to be a slant lightlike submanifold (Theorem (4.3)).

2. Preliminaries

An odd-dimensional semi-Riemannian manifold $\bar{M}$ is said to be an indefinite almost contact metric manifold if there exist structure tensors $(\phi, V, \eta, \bar{g})$, where ϕ is a $(1,1)$ tensor field, V is a vector field called structure vector field, η is a 1-form and $\bar{g}$ is the semi-Riemannian metric on $\bar{M}$ satisfying

(2.1) $\quad \phi^2 X = -X + \eta(X)V, \quad \eta \circ \phi = 0, \quad \phi V = 0, \quad \eta(V) = 1,$

(2.2) $\quad \bar{g}(\phi X, \phi Y) = \bar{g}(X, Y) - \eta(X)\eta(Y) \quad \bar{g}(X, V) = \eta(X),$

for $X, Y \in \Gamma(T\bar{M})$, where $T\bar{M}$ denotes the Lie algebra of vector fields on $\bar{M}$.

An indefinite almost contact metric manifold $\bar{M}$ is called an indefinite Sasakian manifold if (see [11]),

(2.3) $\quad (\bar{\nabla}_X \phi)Y = -\bar{g}(X,Y)V + \epsilon\eta(Y)X, \quad \text{and} \quad \bar{\nabla}_X V = \phi X,$

for any $X, Y \in \Gamma(T\bar{M})$, where $\bar{\nabla}$ denote the Levi-Civita connection on $\bar{M}$ and $\epsilon = \pm 1$.

A submanifold M^m immersed in a semi-Riemannian manifold $(\bar{M}^{m+n}, \bar{g})$ is called an r-lightlike submanifold [6], if it admits a degenerate metric g induced from $\bar{g}$, whose radical distribution $RadTM = TM \cap TM^\perp$ is of rank r, where $0 \leq r \leq min\{m, n\}$. Let $S(TM)$ be a screen distribution which is a semi-Riemannian complementary distribution of $RadTM$ in TM, that is,

(2.4) $\quad\quad\quad TM = RadTM \perp S(TM),$

and $S(TM^\perp)$ be a screen transversal vector bundle, which is a semi-Riemannian complementary vector bundle of $RadTM$ in $TM^\perp$. For any local basis $\{\xi_i\}$ of $RadTM$, there exists a null vector bundle $ltr(TM)$ in $(S(TM))^\perp$ such that $\{N_i\}$ is a basis of $ltr(TM)$ satisfying

(2.5) $\quad\quad\quad \bar{g}(N_i, N_j) = 0 \quad \text{and} \quad \bar{g}(N_i, \xi_j) = \delta_{ij},$

for any $i, j \in \{1, 2, ..., r\}$. Let $tr(TM)$ be the complementary (but not orthogonal) vector bundle to TM in $T\bar{M}|_M$. Then

(2.6) $$tr(TM) = ltr(TM) \perp S(TM^\perp).$$

(2.7)
$$T\bar{M}|_M = TM \oplus tr(TM) = (RadTM \oplus ltr(TM)) \perp S(TM) \perp S(TM^\perp).$$

Let $\bar{\nabla}$ and ∇ denote the linear connections on $\bar{M}$ and M, respectively. Then the Gauss and Weingarten formulae are given by

(2.8) $\quad \bar{\nabla}_X Y = \nabla_X Y + h(X, Y), \quad \bar{\nabla}_X U = -A_U X + \nabla_X^\perp U,$

$X, Y \in \Gamma(TM), U \in \Gamma(tr(TM))$, where $\{\nabla_X Y, A_U X\}$ and $\{h(X, Y), \nabla_X^\perp U\}$ belongs to $\Gamma(TM)$ and $\Gamma(tr(TM))$, respectively. Here ∇ is a torsion-free linear connection on M, h is a symmetric bilinear form on $\Gamma(TM)$ which is called the second fundamental form, A_U is a linear operator on M, known as the shape operator.

Considering the projection morphisms L and S of $tr(TM)$ on $ltr(TM)$ and $S(TM^\perp)$, respectively then using (2.6), Gauss and Weingarten formulae become

(2.9)
$$\bar{\nabla}_X Y = \nabla_X Y + h^l(X, Y) + h^s(X, Y), \quad \bar{\nabla}_X U = -A_U X + D_X^l U + D_X^s U,$$

where we put $h^l(X, Y) = L(h(X, Y)), h^s(X, Y) = S(h(X, Y)), D_X^l U = L(\nabla_X^\perp U), D_X^s U = S(\nabla_X^\perp U)$.

As h^l and h^s are $\Gamma(ltr(TM))$-valued and $\Gamma(S(TM^\perp))$-valued respectively, therefore they are called as the lightlike second fundamental form and the screen second fundamental form on M. In particular, we have

(2.10) $\quad \bar{\nabla}_X N = -A_N X + \nabla_X^l N + D^s(X, N),$

(2.11) $\quad \bar{\nabla}_X W = -A_W X + \nabla_X^s W + D^l(X, W),$

where $X \in \Gamma(TM)$, $N \in \Gamma(ltr(TM))$ and $W \in \Gamma(S(TM^\perp))$. By using (2.6)-(2.7) and (2.9)-(2.11), we obtain

(2.12) $\quad \bar{g}(h^s(X, Y), W) + \bar{g}(Y, D^l(X, W)) = g(A_W X, Y),$

for any $X, Y \in \Gamma(TM)$ and $W \in \Gamma(S(TM^\perp))$. Let $\bar{P}$ is a projection of TM on $S(TM)$ then using the decomposition $TM = RadTM \perp S(TM)$, we can write

(2.13) $\quad \nabla_X \bar{P} Y = \nabla_X^* \bar{P} Y + h^*(X, \bar{P} Y), \quad \nabla_X \xi = -A_\xi^* X + \nabla_X^{*t} \xi,$

for any $X, Y \in \Gamma(TM)$ and $\xi \in \Gamma(RadTM)$, where $\{\nabla_X^* \bar{P}Y, A_\xi^* X\}$ and $\{h^*(X, \bar{P}Y), \nabla_X^{*t} \xi\}$ belong to $\Gamma(S(TM))$ and $\Gamma(RadTM)$, respectively. Here ∇^* and ∇_X^{*t} are linear connections on $S(TM)$ and $RadTM$ respectively. By using (2.9), (2.10) and (2.13), we obtain
(2.14)
$$\bar{g}(h^l(X, \bar{P}Y), \xi) = g(A_\xi^* X, \bar{P}Y), \quad \bar{g}(h^*(X, \bar{P}Y), N) = \bar{g}(A_N X, \bar{P}Y).$$

3. SLANT LIGHTLIKE SUBMANIFOLDS

A lightlike submanifold has two distributions, namely the radical distribution and the screen distribution. The radical distribution is totally lightlike and it is not possible to define angle between two vector fields of the radical distribution where the screen distribution is non-degenerate. There are some definitions for angle between two vector fields in Lorentzian setup [10], but not appropriate for our goal. Therefore to introduce the notion of slant lightlike submanifolds one needs a Riemannian distribution. For such distribution Sahin and Yildirim [13] proved the following lemmas.

Lemma 3.1. *Let M be an r-lightlike submanifold of an indefinite Sasakian manifold $\bar{M}$ of index $2q$. Suppose that $\phi RadTM$ is a distribution on M such that $RadTM \cap \phi RadTM = \{0\}$. Then $\phi ltr(TM)$ is a subbundle of the screen distribution $S(TM)$ and $\phi ltr(TM) \cap \phi RadTM = \{0\}$.*

Lemma 3.2. *Let M be an r-lightlike submanifold of an indefinite Sasakian manifold $\bar{M}$ of index $2r$. Suppose that $\phi RadTM$ is a distribution on M such that $RadTM \cap \phi RadTM = \{0\}$. Then any complementary distribution to $\phi ltr(TM) \oplus \phi(RadTM)$ in screen distribution $S(TM)$ is Riemannian.*

Definition 3.3. *([13]) Let M be an r-lightlike submanifold of an indefinite Sasakian manifold $\bar{M}$ of index $2r$. Then we say that M is a slant lightlike submanifold of $\bar{M}$ if the following conditions are satisfied:*

(A) *$RadTM$ is a distribution on M such that $\phi RadTM \cap RadTM = \{0\}$.*
(B) *For each non zero vector field X tangent to $\bar{D} = D \perp \{V\}$ at $x \in U \subset M$, if X and V are linearly independent, then the angle $\theta(X)$ between ϕX and the vector space $\bar{D}_x$ is constant, that is, it is independent of the choice of $x \in U \subset M$*

and $X \in \bar{D}_x$, where $\bar{D}$ is complementary distribution to $\phi ltr(TM) \oplus \phi Rad TM$ in screen distribution $S(TM)$.

The constant angle $\theta(X)$ is called the slant angle of the distribution $\bar{D}$. A slant lightlike submanifold M is said to be proper if $\bar{D} \neq \{0\}$, and $\theta \neq 0, \frac{\pi}{2}$.

Since a submanifold M is invariant (respectively anti-invariant) if $\phi T_p M \subset T_p M$, (respectively $\phi T_p M \subset T_p M^\perp$), for any $p \in M$. Therefore from above definition, it is clear that M is invariant (respectively anti-invariant) if $\theta(X) = 0$, (respectively $\theta(X) = \frac{\pi}{2}$).

Then the tangent bundle TM of M is decomposed as
(3.1)
$$TM = RadTM \perp S(TM) = RadTM \perp (\phi RadTM \oplus \phi ltr(TM)) \perp \bar{D},$$
where $\bar{D} = D \perp \{V\}$. Therefore for any $X \in \Gamma(TM)$, we write
(3.2)
$$\phi X = TX + FX,$$
where TX is the tangential component of ϕX and FX is the transversal component of ϕX. Similarly for any $U \in \Gamma(tr(TM))$, we write
(3.3)
$$\phi U = BU + CU,$$
where BU is the tangential component of ϕU and CU is the transversal component of ϕU. Using the decomposition in (3.1), we denote by P_1, P_2, Q_1, Q_2 and $\bar{Q}_2$ be the projections on the distributions $RadTM$, $\phi RadTM$, $\phi ltr(TM)$, D and $\bar{D} = D \perp V$, respectively. Then for any $X \in \Gamma(TM)$, we can write
(3.4)
$$X = P_1 X + P_2 X + Q_1 X + \bar{Q}_2 X,$$
where $\bar{Q}_2 X = Q_2 X + \eta(X) V$. Applying ϕ to (3.4), we obtain
(3.5)
$$\phi X = \phi P_1 X + \phi P_2 X + FQ_1 X + TQ_2 X + FQ_2 X.$$
Then using (3.2) and (3.3), we get
(3.6)
$$\phi P_1 X = TP_1 X \in \Gamma(\phi RadTM), \quad \phi P_2 X = TP_2 X \in \Gamma(RadTM),$$
(3.7)
$$FP_1 X = FP_2 X = 0, \quad TQ_2 X \in \Gamma(D), \quad FQ_1 X \in \Gamma(ltr(TM)).$$

Lemma 3.4. *Let M be a slant lightlike submanifold of an indefinite Sasakian manifold $\bar{M}$ then $FQ_2 X \in \Gamma(S(TM^\perp))$, for any $X \in \Gamma(TM)$.*

Proof: Using (2.5) and (2.6) it is clear that $FQ_2X \in \Gamma(S(TM^\perp))$ if $g(FQ_2X, \xi) = 0$, for any $\xi \in \Gamma(Rad(TM))$. Therefore $g(FQ_2X, \xi) = g(\phi Q_2 X - TQ_2X, \xi) = g(\phi Q_2 X, \xi) = -g(Q_2X, \phi\xi) = 0$. Hence the result follows.

Thus from the Lemma (3.4) it follows that $F(D_p)$ is a subspace of $S(TM^\perp)$. Therefore there exists an *invariant* subspace μ_p of $T_p\bar{M}$ such that

(3.8) $$S(T_pM^\perp) = F(D_p) \perp \mu_p,$$

therefore

(3.9) $\quad T_p\bar{M} = S(T_pM) \perp \{Rad(T_pM) \oplus ltr(T_pM)\} \perp \{F(D_p) \perp \mu_p\}.$

Now, differentiating (3.5) and using (2.9)-(2.11), (3.2) and (3.3), for any $X, Y \in \Gamma(TM)$, we have
(3.10)
$(\nabla_X T)Y = A_{FQ_1Y}X + A_{FQ_2Y}X + Bh(X,Y) \quad g(X,Y)V + \epsilon\eta(Y)X,$

and

(3.11) $\quad \begin{aligned} D^s(X, FQ_1Y) + D^l(X, FQ_2Y) &= F\nabla_X Y - h(X, TY) + Ch^s(X,Y) \\ &\quad - \nabla_X^s FQ_2Y - \nabla_X^l FQ_1Y. \end{aligned}$

By using Sasakian property of $\bar{\nabla}$ with (2.8), we have the following lemmas.

Lemma 3.5. *Let M be a slant lightlike submanifold of an indefinite Sasakian manifold $\bar{M}$ then we have*

(3.12) $\quad (\nabla_X T)Y = A_{FY}X + Bh(X,Y) - g(X,Y)V + \epsilon\eta(Y)X,$

(3.13) $\quad\quad\quad (\nabla_X^t F)Y = Ch(X,Y) - h(X,TY),$

where $X, Y \in \Gamma(TM)$ and
(3.14)
$(\nabla_X T)Y = \nabla_X TY - T\nabla_X Y, \quad (\nabla_X^t F)Y = \nabla_X^t FY - F\nabla_X Y.$

In [13], Sahin and Yildirim proved the following theorem.

Theorem 3.6. *Let M be a q lightlike submanifold of an indefinite Sasakian manifold $\bar{M}$. Then M is a slant lightlike submanifold if and only if*

(i) $\phi(RadTM)$ is a distribution on M such that $\phi RadTM \cap Rad(TM) = \{0\}$.

(ii) $\bar{D} = \{X \in \Gamma(\bar{D}) : T^2 X = -\lambda(X - \eta(X)V\}$ is a distribution such that it is complementary to $\phi ltr(TM) \oplus \phi RadTM$, where $\lambda = -cos^2\theta$.

Lemma 3.7. Let M be a slant lightlike submanifold of an indefinite Sasakian manifold $\bar{M}$. Then we have

(3.15) $\quad g(T\bar{Q}_2 X, T\bar{Q}_2 Y) = cos^2\theta[g(\bar{Q}_2 X, \bar{Q}_2 Y) - \eta(\bar{Q}_2 X)\eta(\bar{Q}_2 Y)]$

and

(3.16) $\quad g(F\bar{Q}_2 X, F\bar{Q}_2 Y) = sin^2\theta[g(\bar{Q}_2 X, \bar{Q}_2 Y) - \eta(\bar{Q}_2 X)\eta(\bar{Q}_2 Y)]$

for any $X, Y \in \Gamma(TM)$.

Proof: From (2.1) and (3.2), we obtain

$$g(T\bar{Q}_2 X, T\bar{Q}_2 Y) = -g(\bar{Q}_2 X, T^2 \bar{Q}_2 Y), \quad \forall \ X, Y \in \Gamma(TM).$$

Then from Theorem (3.6), we obtain (29) and (30). This completes the proof.

4. Characterization of Slant lightlike submanifolds of an indefinite Sasakian manifold

Theorem 4.1. Let M be a slant lightlike submanifold of an indefinite Sasakian manifold $\bar{M}$. Then M is anti-invariant lightlike submanifold of $\bar{M}$, if and only if, Q is parallel.

Proof: Let M be a slant lightlike submanifold of an indefinite Sasakian manifold $\bar{M}$, then using part (ii) of Theorem (3.6) for X and Y in TM, we have

(4.1) $\qquad T^2 Y = QY = cos^2\theta(Y - \eta(Y)V),$

this implies

(4.2) $\qquad Q\bar{\nabla}_X Y = cos^2\theta(\bar{\nabla}_X Y - \eta(\bar{\nabla}_X Y)V).$

By taking covariant derivative of (31) with respect to $X \in TM$, we get

(4.3)
$\bar{\nabla}_X QY = cos^2\theta(\bar{\nabla}_X Y - \eta(\bar{\nabla}_X Y)V - g(Y, \bar{\nabla}_X V)V - \eta(Y)\bar{\nabla}_X V).$

Using (32) and (33), we obtain

(4.4) $\qquad (\bar{\nabla}_X Q)Y = \cos^2\theta(g(Y, \nabla_X V)V + \eta(Y)\nabla_X V),$

then further using (2.3), we get

(4.5) $\qquad (\bar{\nabla}_X Q)Y = \cos^2\theta(g(Y, TX)V + \eta(Y)TX).$

Thus the assertion follows from (4.5).

Lemma 4.2. *Let M be an immersed submanifold of an indefinite Sasakian manifold $\bar{M}$ such that V is tangent to M. Then for any $X, Y \in TM$, we have*

(4.6) $\qquad\qquad R(X, Y)V = (\nabla_X T)Y - (\nabla_Y T)X.$

where ∇, R are respectively the Levi-Civita connection and the curvature tensor field associated to the metric induced by $\bar{M}$ on M. Moreover

(4.7) $\qquad\qquad R(V, X)V = QX + \nabla_V TX$

(4.8) $\qquad\qquad R(X, V, X, V) = g(QX, X)$

Proof: Using (2.3) and (2.9), for any $X, Y \in TM$, we obtain

(4.9) $\qquad\qquad \phi X = \nabla_X V + h^l(X, V) + h^s(X, V)$

Using (16) and comparing tangential component of (4.9), we get

(4.10) $\qquad\qquad\qquad TX = \nabla_X V.$

Using (4.10) in (3.14), we have

(4.11) $\qquad (\nabla_X T)Y = \nabla_X TY - T\nabla_X Y = \nabla_X \nabla_Y V - \nabla_{\nabla_X Y} V.$

Similarly

(4.12) $\qquad (\nabla_Y T)X = \nabla_Y TX - T\nabla_Y X = \nabla_Y \nabla_X V - \nabla_{\nabla_Y X} V.$

Thus by subtracting (4.11) and (4.12), we get (4.6). Next, substitute $X = V$ and $Y = X$ in (4.6), we get

(4.13) $\quad R(V,X)V = (\nabla_V T)X - (\nabla_X T)V = (\nabla_V T)X + QX.$

Now taking the scalar product of the above equation with X, and using

(4.14) $\quad g((\nabla_V T)X, X) = g(\nabla_V X, TX) - g(TX, \nabla_V X) = 0,$

we obtain (4.8). This proves the lemma.

Theorem 4.3. *Let M be an immersed submanifold of an indefinite Sasakian manifold $\bar{M}$, such that the characteristic vector field V of $\bar{M}$ is tangent to M. If $\theta \in (0, \frac{\Pi}{2})$, then the following statements are equivalent:*

(a) M *is slant, with slant angle* θ.
(b) *For any $x \in M$, the sectional curvature of any 2-plane of $T_x M$ containing V_x equals $\cos^2\theta$.*

Proof: Let the statement (a) be true, then for any unit vector field $X \perp V$, using Theorem 3.6(ii), we get

$$QX = \cos^2\theta X,$$

then by virtue of (4.8), we yield

(4.15) $\quad\quad\quad\quad R(X, V, X, V) = \cos^2\theta,$

hence (b) is proved. Conversely, assuming that (b) is true, then for any $X \in TM$, we can write

(4.16) $\quad\quad X = (P_1 X + P_2 X + Q_1 X + Q_2 X) + (\eta(X)V).$

this further can be written as

(4.17) $\quad\quad\quad\quad X = X_V^\perp + X_V.$

where $X_V = \eta(X)V$ and $X_V^\perp$ is the component of X perpendicular to the V, then using (4.15) and (4.17), we get

(4.18) $\quad\quad \dfrac{R(X_V^\perp, V, X_V^\perp, V)}{|X_V^\perp|^2} = \cos^2\theta,$

this implies

$$(4.19) \qquad R(X_V^\perp, V, X_V^\perp, V) = \cos^2\theta |X_V^\perp|^2.$$

Let X be a unit vector such that $QX = 0$. Then from (4.6) and (4.19), we have

$$(4.20) \qquad \cos^2\theta |X_V^\perp|^2 = 0$$

Since $\theta \in (0, \frac{\pi}{2})$ therefore $\cos\theta \neq 0$, hence $X_V^\perp = 0$. Then from (4.17) $X = X_V$. This implies that, at each point $x \in M$, we have

$$(4.21) \qquad Ker(Q) = \langle X_V \rangle$$

Let A be the matrix of the endomorphism Q at $x \in M$, then for a unit vector field X on M, $QX = AX$. Since $Q(X_V) = 0$ and $X = X_V$. Then using (4.8) and (4.19), we get

$$(4.22) \qquad A = \cos^2\theta I$$

Choosing $\lambda = \cos^2\theta$, then using (4.21) and Theorem 3.6 (ii), M is slant lightlike submanifold in $\bar{M}$ with slant angle θ. Finally, suppose that $\cos\theta = 0$ and X is an arbitrary unit vector field such that $QX = \lambda X$ where $\lambda \in C^\infty(M)$. Then, from (4.8) and (4.19), we infer $g(QX, X) = 0$, that is, $\lambda = 0$ and therefore $Q = 0$, which means M is an anti-invariant lightlike submanifold. Hence the proof.

REFERENCES

[1] C. L. Bejan and K. L. Duggal, Global lightlike manifolds and harmonicity, Kodai Math. J., **28** (2005), 131-145.

[2] J. L. Cabrerizo, A. Carriazo, L. M. Fernandez, and M. Fernandez, Semi-slant submanifolds of a Sasakian manifold, Geom. Dedicata, **78** (1999), 183-199.

[3] J. L. Cabrerizo, A. Carriazo, L. M. Fernandez, and M. Fernandez, Slant submanifolds in Sasakian manifolds, Glasg. Math. J., **42** (2000), 125-138.

[4] B. Y. Chen, Slant immersions, Bull. Austral. Math. Soc., **41** (1990), 135-147.

[5] B. Y. Chen, Geometry of Slant submanifolds, Katholieke Universiteit, Leuven, 1990.

[6] K. L. Duggal and A. Bejancu, Lightlike submanifolds of semi-Riemannian manifolds and applications, Vol. 364 of Mathematics and its Applications, Kluwer Academic Publishers, Dordrecht, The Netherlands, 1996.

[7] K. L. Duggal and B. Sahin, Screen Cauchy Riemann lightlike submanifolds, Acta Math. Hungar., **106** (2005), 125-153.

[8] M. A. Khan, K. Singh and V. A. Khan, Slant submanifolds of LP- contact manifolds, Differ. Geom. Dyn. Syst., **12** (2010), 102-108.

[9] A. Lotta, Slant submanifolds in contact geometry, Bull. Soc. Sci. Math. Roumanie, **39** (1996), 183-198.

[10] B. O'Neill, Semi-Riemannian geometry with applications to relativity, Academic Press, New York, 1983.

[11] Rakesh Kumar, Rachna Rani and R. K. Nagaich, On sectional curvature of ϵ- Sasakian manifolds, Int. J. Math. Math. Sci., Vol (2007), Article ID 93562, 8 pages.

[12] B. Sahin, Slant lightlike submanifolds of indefinite Hermitian manifolds, Balkan J. Geom. Appl., **13**(2008), 107-119.

[13] B. Sahin and C. Yildirim, Slant lightlike submanifolds of indefinite Sasakian manifolds, Filomat, **26** (2012), 71-81.

[14] K. Yano and M. Kon, Stuctures on manifolds, Vol. 3 of Series in Pure Mathematics, World Scientific, Singapore, 1984.

Rashmi
School of Mathematics and Computer Applications,
Thapar University, Patiala, India.
E-mail: rashmi.sachdeva86@gmail.com

Rakesh Kumar
University College of Engineering, Punjabi University, Patiala, India.
E-mail: dr_rk37c@yahoo.co.in
and

S. S. Bhatia
School of Mathematics and Computer Applications,
Thapar University, Patiala, India.
E-mail: ssbhatia@thapar.edu

Differential Geometry, Functional Analysis and Applications
Editors: Mohammad Hasan Shahid, Sharfuddin Ahmad *et al.*
Copyright © 2015, Narosa Publishing House, New Delhi

ON DOUBLY TWISTED PRODUCT CR-SUBMANIFOLDS

MAJID ALI CHOUDHARY AND MOHAMMED JAMALI

ABSTRACT. *A. Bejancu [2] defined and studied CR-submanifolds of a Kaehler manifold. Since then many papers appeared on this topic. In {[5],[4]} B. Y. Chen and in [11] B. Sahin studied warped product submanifolds of a Kaehlerian manifold. In the present note, we investigate the existence of doubly twisted product CR-submanifolds in locally conformal quaternion Kachler manifolds and prove that there do not exist doubly twisted product CR- submanifolds in locally conformal quaternion Kaehler manifolds.*

1. INTRODUCTION

R. L. Bishop and B. O'Neill in [3] tossed the concept of warped product manifolds while constructing example of Riemannian manifolds with negative sectional curvatures. In general, doubly twisted product manifolds can be considered as generalization of warped products. Warped product manifolds have importance as they are widely used to provide setting to model space time near black holes or bodies with large gravitational force. Suppose that (B, g_B) and (F, g_F) be semi-Riemannian manifolds of dimensions m and n, respectively and further suppose that $\pi : B \times F \to B$ and $\sigma : B \times F \to F$ be the canonical projections. Let $b : B \times F \to (0, \infty)$, $f : B \times F \to (0, \infty)$ be smooth functions. Then the doubly twisted product ([7],[10]) of (B, g_B) and (F, g_F) with twisting functions b and f is defined to be the product manifold $M = B \times F$ with metric tensor $g = f^2 g_B \oplus b^2 g_F$. Denoting this kind of manifolds by $_f B \times_b F$ and by $F(B)$ denoting the algebra of smooth functions on B and by $\Gamma(E)$ the $F(B)$ module

---

2000 *Mathematics Subject Classification.* 53C25, 53C40, 53C42.
Key words and phrases. Doubly twisted warped product, l.c.q.K. manifold.

of smooth sections of a vector bundle E (same notation for any other bundle) over B. If $X \in \Gamma(TB)$ and $V \in \Gamma(TF)$, then from Proposition 1 of [7], we have

$$(1.1) \qquad \nabla_X V = V(\ln f)X + X(\ln b)V$$

where ∇ denotes the Levi-Civita connection of the doubly twisted product $_fB \times_b F$ of (B, g_B) and (F, g_F). In particular, if $f = 1$, then $B \times_b F$ is called the twisted product of (B, g_B) and (F, g_F) with twisting function b. We note that the notion of twisted products was introduced in [5]. If $M = B \times_b F$ is a twisted product manifold, then (1.1) becomes

$$(1.2) \qquad \nabla_X V = X(\ln b)V$$

On the other hand, locally conformal Kaehler manifold was introduced by I. Vaisman in [13]. However, the geometry of the locally conformal quaternion Kaehler manifolds has been studied in ([6],[8],[9]) and their QR-Submanifolds have been studied in [12].

A locally conformal quaternion Kaehler manifold (Shortly, l.c.q.K manifold) is a quaternion Hermitian manifold whose metric is conformal to a quaternion Kaehler metric in some neighborhood of each point. The main difference between l.c.K. manifold and l.c.q.K. manifold is that the Lee form of a compact l.c.q.K. manifold can be chosen as parallel form without any restrictions [6].

A. Bejancu [2] defined and studied CR-submanifolds of a Kaehler manifold. B. Y. Chen [4] introduced twisted product CR-submanifolds in Kaehler manifolds and showed that a twisted product CR-submanifold in the form $M_\perp \times_f M_T$ is a CR-product. He also considered twisted product CR- submanifolds in the form $M_T \times_f M_\perp$ and established a general sharp inequality for twisted product CR-submanifolds in Kaehler manifolds. Then, B. Sahin [11] studied doubly warped product and doubly twisted warped product submanifolds of a Kaehlerian manifold.

In this note, we investigate the existence of doubly twisted product CR-submanifolds in locally conformal quaternion Kaehler manifolds and give a result showing that there do not exist doubly twisted product CR-submanifolds in locally conformal quaternion Kaehler manifolds.

2. Preliminaries

In this section, we recall the definitions of l.c.q.K. manifold and CR-submanifold.

Definition 2.1. *Let $(\widetilde{M}, J, g)$ be a quaternion Hermitian manifold where H is a sub-bundle of $End(T\widetilde{M})$ of rank 3 which is spanned by almost complex structures J_1, J_2, and J_3. The quaternion Hermitian metric g is said to be a quaternion Kaehler metric if its Levi-Civita connection $\widetilde{\nabla}$ satisfies $\widetilde{\nabla} H \subset H$.*

A quaternion Hermitian manifold with metric g is called a locally conformal quaternion Kaehler (l.c.q.K.) manifold if over neighborhoods U_i covering $\widetilde{M}$, $g\mid_{U_i} = e^{f_i} g_i$ where g_i is a quaternion Kaehler metric on U_i. In this case, the Lee form ω is locally defined by $\omega\mid_{U_i} = df_i$ and satisfies [8]

$$d\theta = \omega \wedge \theta, \quad d\omega = 0.$$

Let $\widetilde{M}$ be l.c.q.K. manifold and $\widetilde{\nabla}$ denotes the Levi Civita connection of $\widetilde{M}$. Let B be the Lee vector field given by $g(X, B) = \omega(X)$. Then for l.c.q.K. manifold we have [8]

$$(\widetilde{\nabla}_X J_a)Y = \tfrac{1}{2}\{\theta(Y)X - \omega(Y)J_a X - g(X,Y)A - \Omega(X,Y)B\}$$
(2.1) $$+ Q_{ab}(X)J_b Y + Q_{ac}(X)J_c Y$$

for any $X, Y \in T\widetilde{M}$, where Q_{ab} is a skew symmetric matrix of local forms, $\theta = \omega o J_a$ and $A = -J_a B$.

A Riemannian manifold M, isometrically immersed in a l.c.q.K. manifold $\widetilde{M}$ is called CR-submanifold [1] if there exists on M a differentiable holomorphic distribution D, i.e. $J_a D = D$ for $a = 1, 2, 3$ whose orthogonal complement $D^\perp$ of D in $T(M)$ is totally real distribution on M, i.e. $J_a D^\perp \subset T(M)^\perp$ for $a = 1, 2, 3$. A CR-submanifold is called holomorphic submanifold if $dim\ D^\perp = 0$, totally real if $dim\ D = 0$ and proper if it is neither holomorphic nor totally real.

Let M be a Riemannian manifold isometrically immersed in $\widetilde{M}$ and denote by the same symbol g the Riemannian metric induced on M. Let TM be the Lie algebra of vector fields in M and $TM^\perp$, the set of all vector fields normal to M. Denote by ∇ the Levi-Civita connection of M. Then the Gauss and Weingarten formulas are given by

(2.2) $$\widetilde{\nabla}_X Y = \nabla_X Y + h(X,Y)$$

and

(2.3) $$\widetilde{\nabla}_X N = -A_N X + \nabla_x^\perp N$$

for any $X, Y \in TM$ and any $N \in TM^\perp$, where $\nabla^\perp$ is the connection in the normal bundle $TM^\perp$, h is the second fundamental form of M and A_N is the Weingarten endomorphism associated with N. The second fundamental form and the shape operator A are related by

(2.4) $$g(A_N X, Y) = g(h(X,Y), N)$$

3. Doubly twisted Product CR-submanifolds

In this section, we consider CR-submanifolds which are doubly twisted products in the form $_f M_T \times_b M_\perp$, where M_T is a holomorphic submanifold and $M_\perp$ is a totally real submanifold of $\widetilde{M}$.

Theorem 3.1. *Let $\widetilde{M}$ be a locally conformal quaternion Kaehler manifold. Then there do not exist doubly twisted product CR-submanifolds of $\widetilde{M}$ which are not (singly) twisted product CR-submanifolds in the form $_f M_T \times_b M_\perp$ such that M_T is a holomorphic submanifold and $M_\perp$ is a totally real submanifold of $\widetilde{M}$.*

Proof. Let us suppose that M be a doubly twisted product CR-submanifold of a l.c.q.K. manifold $\widetilde{M}$. Then from (2.1) we have

$$(\widetilde{\nabla}_X J_a)Y = \tfrac{1}{2}\{\theta(Y)X - \omega(Y)J_a X - g(X,Y)A - \Omega(X,Y)B\}$$
$$+ Q_{ab}(X)J_b Y + Q_{ac}(X)J_c Y$$

for $X, Y, B \in \Gamma(D)$ and $V \in \Gamma(D^\perp)$. Using (2.2) we have

(3.1) $$g(\nabla_X Y, V) = g(h(X, J_a Y), J_a V)$$

From equation (1.1), we get

$$g(\nabla_X V, Y) = V(lnf)g(X, Y)$$

Taking into account that D and $D^\perp$ are orthogonal, we obtain

(3.2) $$-g(V, \nabla_X Y) = V(lnf)g(X, Y)$$

So, from (3.1) and (3.2), we have

(3.3) $$-g(h(X, J_a Y), J_a V) = V(lnf)g(X, Y)$$

for $X, Y \in \Gamma(D)$ and $V \in \Gamma(D^\perp)$.

But, by the use of Gauss formula we have,

$$g(h(X, J_a Y), J_a V) = g(\widetilde{\nabla}_{J_a Y} X, J_a V)$$

which upon applying equation (2.1) reduces to

$$g(h(X, J_a Y), J_a V) = -g(\widetilde{\nabla}_{J_a Y} J_a X, V).$$

In the light of Gauss formula and taking into account the orthogonality of D and $D^\perp$, above equation reduces to

$$g(h(X, J_a Y), J_a V) = g(\nabla_{J_a Y} V, J_a X)$$

In view of equation (1.1), above equation gives

(3.4) $$g(h(X, J_a Y), J_a V) = V(lnf)g(J_a X, J_a Y)$$

for $X, Y \in \Gamma(D)$ and $V \in \Gamma(D^\perp)$. From equations (3.3) and (3.4), we have $V(lnf)g(X, Y) = 0$. Since D is Riemannian, we get $V(lnf) = 0$. This implies that f only depends on the point of M_T. Thus, we can write

$$g = \overline{g}_{M_T} \oplus b^2 g_{M_\perp}$$

where $\overline{g}_{M_T} = f^2 g_{M_T}$.

Thus, it follows that M is a twisted product CR-submanifold in the form $M_T \times_b M_\perp$ ([4] for twisted product CR-submanifolds). Hence, we conclude that there are no doubly twisted product CR-submanifolds in l.c.q.K. manifold, other than twisted product CR-submanifolds. □

Acknowledgements. The first author is thankful to Department of Science and Technology, Government of India, for its financial assistance provided through Inspire Fellowship No. DST/INSPIRE Fellowship/2009/[xxv] to carry out this research work.

References

[1] Bejancu, A., Geometry of CR-Submanifolds, *Kluwer Academic Pub., Dortrecht*, (1986).

[2] Bejancu, A., CR-Submanifolds of a Kaehler Manifold, I-II, *Proc. Amer. Math. Soc.,* **69**(1978),135-142, *Trans.Amer.Math.Soc.,* **250** (1979),333-345.

[3] Bishop, R.L., Neill B. O'., Manifolds of negative curvature, *Trans. Amer. Math. Soc.,* **145**(1969), 1-49.

[4] Chen, B. Y., Twisted product CR-submanifolds in Kaehler manifolds, *Tamsui Oxf. J. Math. Sci.,***16, 2** (2000), 105-121.

[5] Chen, B. Y., Geometry of warped product CR-submanifolds in Kaehler manifold, *Monatsh. Math,* **133**(2001), 177-195.

[6] Dragomir, S., Ornea, L., Locally Conformal Kahler Geometry. *Basel: Birkhauser,* (1998).

[7] Lopez M. F., Garcia-Rio E., D. N. Kupeli, B. Unal, A curvature condition for a twisted product to be a warped product,*Manuscripta Math.,***106**, (2001), 213-217.

[8] Ornea, L., Piccini, P., Locally Conformal Kaehler Structures in Quaternionic Geometry, *Trans. Amer. Math. Soc.,***349, No. 2** (1997), 641-645.

[9] Ornea, L., Weyl Structures on Quaternionic Manifolds. A State of the Art, *arXiv:math.* **DG/0105041, v1,** (2002).

[10] Ponge R., Reckziegel H., Twisted products in Pseudo-Riemannian Geometry,*Geometriae Dedicata,***48**, (1993), 15-25.

[11] Sahin, B., Notes on doubly warped and doubly twisted product CR-submanifolds of Kaehler manifolds, *Matematiqki Bechnk,* **59** (2007), 205-210.

[12] Sahin, B., Gunes, R., QR-Submanifolds of a Locally Conformal Quaternion Kaehler Manifolds,*Publ.Math.Debrecen,***63** (2003).

[13] Vaisman, I., On Locally Conformal Almost Kaehler Manifolds, *Israel J. Math.,***24** (1976), 338-351.

MAJID ALI CHOUDHARY, DEPARTMENT OF MATHEMATICS, JAMIA MILLIA ISLAMIA, NEW DELHI -110025, INDIA
E-mail address: `majid_alichoudhary@yahoo.co.in`

MOHAMMED JAMALI, DEPARTMENT OF APPLIED SCIENCES AND HUMANITIES, AL-FALAH UNIVERSITY, DHAUJ, FARIDABAAD, HARYANA, INDIA
E-mail address: `jamali_dbd@yahoo.co.in`

Differential Geometry, Functional Analysis and Applications
Editors: Mohammad Hasan Shahid, Sharfuddin Ahmad *et al.*
Copyright © 2015, Narosa Publishing House, New Delhi

A NOTE ON TRANS-SASAKAN MANIFOLDS

RAJENDRA PRASAD, KWANG-SOON PARK, AND JAI PRAKASH

ABSTRACT. In this paper we have studied conformally flat and quasi-conformally flat trans-Sasakian manifolds. We have proved that in a conformally flat trans-Sasakian manifold, ξ is an eigen vector of Ricci operator Q. Some expressions, lemmas and theorems for trans-Sasakian manifolds have been explored. For trans-Sasakian manifolds expression for $\phi Q - Q\phi$ has been found. It is shown that Ricci operator Q does not commute with $(1,1)$ tensor field ϕ for such manifolds. If a trans-Sasakian manifold is conformally flat then it is η−Einstein.

1. INTRODUCTION

In [3], Okumura showed that a conformally flat Sasakian manifold of dimension > 3 is of constant curvature and in [4], Tanno extended this result to the K-contact case and for dimension ≥ 3. Some new examples of conformally flat manifolds, as a step towards a classification of such manifolds upto conformal equivalence was given by Kulkarni in 1972 [8]. It is known that the Ricci operator Q commute with $(1,1)$ tensor field ϕ for a Sasakian manifold and Kenmotsu manifold but this commutativity need not hold for a contact metric manifold and an almost contact metric manifold. It is shown in this paper that above commutativity does not hold for a trans-sasakian manifold also.

In [1], Blair proved that there are no contact metric manifolds of vanishing curvature and of dimension ≥ 5. Generalizing this result Olszak [2] proved that any contact manifold of constant sectional curvature and of dimension≥ 5 has the sectional curvature equal to

---

2000 *Mathematics Subject Classification.* Primary 05C38, 15A15; Secondary 05A15, 15A18.

Key words and phrases. Trans-Sasakian manifolds, Conformally flat manifolds, Ricci operator, Ricci tensor.

1 and is a Sasakian manifold. Olszak also proved that on a conformally flat contact metric manifold of dimension $2n+1$ $(n > 1)$, the scalar curvature r satisfies $r \leq 2n(2n+1)$, where equality holds if and only if the manifold is Sasakian. In [7], Gosh and Sharma proved the following result.

Theorem: Let M be a $(2n+1)$ dimensional $(n > 1)$ contact strongly pseudo-convex integrable CR manifold such that ξ is an eigen vector of Ricci-operator at each point. If M is conformally flat. then it is of constant curvature 1.

The notion of the quasi-conformal curvature tensor was given by Yanno and Sawaki [10] in 1968. According to them a quasi-conformal curvature tensor $\check{C}$ of $(2n+1)$ dimensional manifold M is defined as

$$\check{C}(X,Y)Z = aR(X,Y)Z + b[S(Y,Z)X - S(X,Z)Y \qquad (1.1)$$
$$+ g(Y,Z)QX - g(X,Z)QY]$$
$$- \frac{r}{(2n+1)}(\frac{a}{2n} + 2b)[g(Y,Z)X - g(X,Z)Y],$$

where a and b are constants and R, S, Q and r are the Riemannian curvature-tensor, the Ricci- tensor, the Ricci operator and the scalar curvature tensor of the manifold respectively. If $a = 1$ and $b = -\frac{1}{2n-1}$, then $\check{C}$ becomes a conformal curvature tensor C, given by

$$C(X,Y)Z = R(X,Y)Z - \frac{1}{2n-1}\{g(Y,Z)QX - g(X,Z)QY\}$$
$$+ S(Y,Z)X - S(X,Z)Y\}$$
$$+ \frac{r}{2n(2n-1)}\{g(Y,Z)X - g(X,Z)Y\}, \qquad (1.2)$$

Let M be a $(2n+1)$- dimensional almost contact metric manifold [10] with almost contact metric structure (ϕ, ξ, η, g), Let M be a $(2n+1)$- dimensional almost contact metric manifold [10] with almost contact metric structure (ϕ, ξ, η, g), where ϕ is a $(1,1)$ tensor field, ξ is a vector field, η is a 1-form and g is a compatible Riemannian metric on M such that

$$\phi^2 = -I + \eta \otimes \xi, \qquad \eta(\xi) = 1, \qquad \phi\xi = 0, \qquad (1.3)$$

$$g(\phi X, \phi Y) = g(X, Y) - \eta(X)\eta(Y), \qquad (1.4)$$

$$g(\phi X, Y) = -g(X, \phi Y), \quad g(X, \xi) = \eta(X), \qquad (1.5)$$

for all $X, Y \in TM$.

An almost contact metric manifold M is called trans-Sasakian manifold if

$$(\nabla_X \phi)Y = \alpha\{g(X,Y)\xi - \eta(Y)X\} + \beta\{g(\phi X, Y)\xi - \eta(Y)\phi X\}, \qquad (1.6)$$

where ∇ is Levi-Civita connection of Riemannian metric g and α, β are smooth functions on M.

From equation(1.7) and equations(1.3), (1.4) and (1.5), we get

$$\nabla_X \xi = -\alpha \phi X + \beta [X - \eta(X)\xi], \qquad (1.7)$$

$$(\nabla_X \eta)Y = -\alpha g(\phi X, Y) + \beta g(\phi X, \phi Y). \qquad (1.8)$$

$$\nabla_\xi \phi = 0 \qquad (1.9)$$

Let us define the tensor h by $2h = \pounds_\xi \phi$, where $\pounds$ is the Lie diferentiation operator.

Further, on such a trans–Sasakian manifold M of dimension $(2n+1)$ with almost contact structure(ϕ, ξ, η, g), the following relations hold [11],

$$S(X,\xi) = \left(2n\left(\alpha^2 - \beta^2\right) - \xi\beta\right)\eta(X) - (2n-1)X\beta - (\phi X)\alpha, \qquad (1.10)$$

$$Q\xi = \left(2n\left(\alpha^2 - \beta^2\right) - \xi\beta\right)\xi - (2n-1)\,grad\beta + \phi\,(grad\alpha)\,. \qquad (1.11)$$

2. Quasi-conformaly Flat and Conformally Flat Manifolds

Let M be a $(2n+1)$–dimensional quasi-conformally flat manifold, then from equation(1.1), we have

$$aR(X,Y)Z = b[-g(Y,Z)QX + g(X,Z)QY \quad (2.1)$$
$$-S(Y,Z)X + S(X,Z)Y]$$
$$+\frac{r}{2n(2n-1)}[\frac{a}{2n} + 2b]$$
$$[g(Y,Z)X - g(X,Z)Y],$$

Let $\{e_1, e_2 ... e_{2n}, e_{2n+1} = \xi\}$ is orthonormal base field. Putting $Y = Z = e_i$ in equation (2.1), we get

$$S(X,W) = \frac{r}{2n+1}g(X,W) \qquad if\ a+(2n-1)b \neq 0 \quad (2.2)$$

Hence we have the following lemma

Lemma 2.1 A $(2n+1)$ $-dimensional$ quasi-conformally flat manifold M is an Einstein manifold if $a+(2n-1)b \neq 0$.

If $a+(2n-1)b - 0$, then from equations(1.1) and(1.2), we have

$$\check{C}(X,Y)Z = a(C(X,Y)Z) \qquad (2.3)$$

If $a = 1, b = -\frac{1}{2n-1}$ then $a+(2n-1)b = 1+(2n-1)\left(-\frac{1}{2n-1}\right) = 0$ so condition $a+(2n-1)b = 0$ is satisfied for conformal curvature tensor.

From equation (2.3), we have following Lemma:

Lemma 2.2 A $(2n+1)$ dimensional quasi-conformally flat manifold M is conformaly flat if $a + (2n-1)b = 0$ and $a \neq 0$.

from equation (2.1) and (2.2), we have

$$R(X,Y)Z = \frac{r(1+4nb/a)}{2n(2n+1)}\{g(Y,Z)X - g(X,Z)Y\},$$
$$if\ a \neq 0,\ a+(2n-1)b \neq 0 \qquad (2.4)$$

Hence we have following theorem

Theorem 2.3 A quasi-conformally flat manifold is a manifold of constant curvature if $a \neq 0, a+(2n-1)b \neq 0$.

Note: A conformally flat almost constant manifold is not in general a manifold of constant curvature.

3. Quasi-conformally Flat Trans-sasakian Manifold

For a trans-Sasakian manifold, we have

$$S(X,\xi) = \left(2n\left(\alpha^2 - \beta^2\right) - \xi\beta\right)\eta(X) - (2n-1)X\beta - (\phi X)\alpha, \tag{3.1}$$

$$S(\xi,\xi) = 2n\left(\alpha^2 - \beta^2 - \xi\beta\right), \tag{3.2}$$

If a trans-Sasakian manifold is quasi-conformally, then from equation (2.2) and (3.2), we have

$$r = 2n(2n+1)\left(\alpha^2 - \beta^2 - \xi\beta\right) \tag{3.3}$$

if $a + (2n+1)b \neq 0$
from equation (2.5), we have

$$\begin{aligned}R(X,Y)Z &= \left(\alpha^2 - \beta^2 - \xi\beta\right)(1 + 4nb/a)\{g(Y,Z)X \\ &\quad - g(X,Z)Y\} \; if \, a + (2n-1)b \neq 0, a \neq 0\end{aligned} \tag{3.4}$$

Lemma 3.2 In a trans-Sasakian manifold M, $h = 0$ [11].
Proof: Using definition of h

$$hX = \frac{1}{2}\left[(\pounds_\xi\phi)X\right] \tag{3.5}$$

$$= \frac{1}{2}\{(\nabla_\xi\phi)X - \nabla_{\phi X}\xi + \phi(\nabla_X\xi)\} \tag{3.6}$$

using equation (1.8) and (1.10), we have
$hX = 0$, for all X

$$h = 0 \tag{3.7}$$

hence we have,

$$h\phi = \phi h$$

Lemma 3.3 In s trans-Sasakian manifold M, the following relation holds:

$$R(X,Y)\phi Z - \phi R(X,Y)Z = (\alpha^2 - \beta^2)[g(X,Z)\phi Y - g(Y,Z)\phi X$$
$$+ g(\phi X, Z) - g(\phi Y, Z) X]$$

$$+ 2\alpha\beta[g(Y,Z)X - g(X,Z)Y$$

$$+ g(\phi X, Z)\phi Y - g(\phi Y, Z)\phi X]$$

$$+ (X\alpha)[g(Y,Z)\xi - \eta(Z)Y]$$

$$- (Y\alpha)[g(X,Z)\xi - \eta(Z)X]$$

$$- (Y\beta)[g(\phi X, Z)\xi - \eta(Z)\phi X] \quad (3.8)$$

for all vector fields X, Y and Z on M.

Proof: We know that

$$R(X,Y)\phi Z - \phi R(X,Y)Z = (\nabla_X \nabla_Y \phi)Z - (\nabla_Y \nabla_X \phi)Z$$
$$- (\nabla_{[X,Y]}\phi)Z \quad (3.9)$$

For trans-Ssakian manifold we have $(\nabla_X \nabla_Y \phi)Z$, $(\nabla_Y \nabla_X \phi)Z$, $(\nabla_{[X,Y]}\phi)Z$

From the equations (3.9), we get the Lemma.

Lemma 3.4 In a conformally flat trans-Sasakian manifold, ξ is an eigen vector of Ricci operator Q.

Proof:

Putting $X = \xi$ in equation (3.8), we get

$$R(\xi, Y)\phi Z - \phi R(\xi, Y)Z = -(\alpha^2 - \beta^2$$
$$-\xi\beta)\{g(\phi Y, Z)\xi - \eta(Z)\phi Y\} \quad (3.10)$$

Let M be a trans-Sasakian manifold of dimension $(2n+1)$

Let $\{e_1, e_2 ... e_{2n+1} = \xi\}$ be orthnormal basis. Putting $Y = Z = e_i$ in equation (3.10) and taking summation, we get

$$\sum_{i=1}^{2n+1} R(\xi, e_i) \phi e_i = \phi Q \xi \qquad (3.11)$$

If the manifold is conformally flat, $C = 0$ then

$$\begin{aligned} R(X,Y)Z &= \frac{1}{(2n-1)}[g(Y,Z)QX - g(X,Z)QY \\ &\quad - S(Y,Z)X + S(X,Z)Y] \\ &\quad + \frac{r}{2n(2n-1)}[g(Y,Z)X - g(X,Z)Y], (3.12)\end{aligned}$$

Putting $X = \xi, Y = e_i, Z = \phi e_i$ in above equation and taking summation in above
using equation(3.8), we get

$$\phi Q\xi = \frac{1}{2n-1}\{(trQ\phi)\xi + \phi Q\xi\} \qquad (3.13)$$

Since $trQ\phi = 0$. Also $n > 1$, therefore $\phi Q\xi = 0$ and hence $Q\xi = (trl)\xi$.
Hence the lemma.
On almost contact metric manifold, define operator l by

$$lX = R(X,\xi)\xi \text{ for every } X \qquad (3.14)$$

obviously $l\xi = 0$
For trans-Sasakian manifold M of dimension $(2n+1)$

$$trl = \sum_{i=1}^{2n+1} g(R(e_i,\xi)\xi, e_i) \qquad (3.15)$$

$$S(X,W) = \frac{r}{2n+1}g(X,W) \qquad if\ a + (2n-1)b \neq 0$$
$$= 2n\left(\alpha^2 - \beta^2 - \xi\beta\right) \qquad (3.16)$$

$$lX = R(X,\xi)\xi = \left(\alpha^2 - \beta^2 - \xi\beta\right)[X - \eta(X)\xi] \qquad (3.17)$$

$$l\phi X = \left(\alpha^2 - \beta^2 - \xi\beta\right)\phi X \qquad (3.18)$$

so, $\phi l = l\phi$ in a trans-Sasakian manifold.
$$\phi l \phi X - lX = \left(\alpha^2 - \beta^2 - \xi\beta\right).0 = 0 \tag{3.19}$$

$$\phi l \phi = l \tag{3.20}$$

so
$$\eta(lX) = g\left(R(X,\xi)\xi,\xi\right) = 0 \tag{3.21}$$

so
$$\eta o l = 0 \tag{3.22}$$

$$K(\xi, X) =' R(\xi, X, \xi, X) \tag{3.23}$$

$$= g\left(R(\xi, X)\xi, X\right) \tag{3.24}$$

$$= -\left(\alpha^2 - \beta^2 - \xi\beta\right)\left[g(X,X) - \eta(X)\eta(X)\right] \tag{3.25}$$

$$K(\xi, \phi X) = -\left(\alpha^2 - \beta^2 - \xi\beta\right)\left[g(\phi X, \phi X)\right]$$

$$= -\left(\alpha^2 - \beta^2 - \xi\beta\right)\left[g(X,X) - \eta(X)\eta(X)\right] \tag{3.26}$$

$$K(\xi, X) = K(\xi, \phi X) \tag{3.27}$$

Lemma 3.5 A conformally flat trans-Sasakian manifold is η- Einstein.

Proof:

For conformally flat trans-Sasakian manifold M of dimension $(2n+1)$, we have $C(X,Y)Z = 0$

$$\begin{aligned}R(X,Y)Z &= -\frac{1}{2n-1}\{g(Y,Z)QX - g(X,Z)QY \\ &\quad + S(Y,Z)X - S(X,Z)Y\} \\ &\quad - \frac{r}{2n(2n-1)}\{g(Y,Z)X - g(X,Z)Y\}\end{aligned} \tag{3.28}$$

Putting $Y = Z = \xi$, we get
So manifold is η−Einstein and hence

$$Q\phi = \phi Q.$$

Differentiating $Q\xi = (trl)\xi$, along an arbitrary vector field X

$$\begin{aligned}(\nabla_X Q)\xi &= Q\{\alpha\phi X + \beta(\eta(X)\xi - X)\} + X(trl)\xi \quad (3.29)\\ &\quad - (trl)\{\alpha\phi X + \beta(\eta(X)\xi - X)\}\end{aligned}$$

$$\begin{aligned}&= \alpha Q\phi X + \beta\eta(X)Q\xi - \beta QX + X(trl)\xi \quad (3.30)\\ &\quad - (trl)\alpha\phi X - \beta(trl)\{\eta(X) - X\}\end{aligned}$$

As $C = 0$, we have $divC = 0$ or equivalently

$$g((\nabla_X Q)Y, Z) - g((\nabla_Y Q)X, Z) = \frac{1}{4n}\{Xrg(Y,Z) - (Yr)g(X,Z)\} \quad (3.31)$$

Putting $Y = Z = \xi$, we get

$$X(trl) - \frac{1}{4n}(Xr) = \{\xi(trl) - \frac{1}{4n}\xi r\}\eta(X) \quad (3.32)$$

First applying exterior derivative on(3.41), using the Poincare lemma: $d^2 = 0$, and then replacing X, Y respectively by $\phi X, \phi Y$ in resulting equation, we get

$$\xi(trl) = \frac{1}{4}(\xi r) \quad (3.33)$$

and hence

$$Xtrl = \frac{1}{4n}Xr \quad (3.34)$$

$$\nabla_X\left(trl - \frac{1}{4n}r\right) = 0 \quad (3.35)$$

for every X

$\left(trl - \frac{1}{4n}r\right)$ is constant.

so

$$r = 4ntrl + const$$

Hence we have following Lemma:

Lemma 3.6 In a conformally flat trans-Sasakian manifold M of dimension $(2n+1)$, the scalar curvature is given by
$r = 4ntrl + const.$

Theorem 3.7 *In a trans-Sasakian manifold M of dimension $(2n+1)$, the following relation holds*

$$\begin{aligned}Q\phi X - \phi QX &= 8\alpha\beta(n-1)\phi^2 X + (X\alpha)\xi - (\xi\alpha)\eta(X)\xi \\ &\quad - (2n-1)(\phi X)\beta\xi + \eta(X) \\ &\quad \phi[(2n-1)\operatorname{grad}\beta - \phi(\operatorname{grad}\alpha)] \end{aligned} \quad (3.36)$$

Proof: Let $\{X_i, \phi X_i, \xi\}$ $(i = 1, 2, ..., n)$ be a local ϕ-basis at any point of the manifold. Then putting $Y = Z = X_i$ in (3.45) and taking summation over i, we obtain by virtue of $\eta(X_i) = 0$, we get

$$\begin{aligned}-\phi Q(\phi X) &= QX - [2n(\alpha^2 - \beta^2) - \xi\beta]\eta(X)\xi \\ &\quad + 8\alpha\beta(n-1)\phi X \\ &\quad + [(2n-1)(\phi X)\alpha - \phi X\beta]\xi \\ &\quad + \eta(X)[(2n-1)\operatorname{grad}\beta - \phi(\operatorname{grad}\alpha)] \end{aligned} \quad (3.37)$$

operating ϕ on both sides, we get

$$Q\phi X - \phi QX \quad (3.38)$$
$$= 8\alpha\beta(n-1)\phi^2 X + (X\alpha)\xi - (\xi\alpha)\eta(X)\xi$$
$$- (2n-1)(\phi X)\beta\xi + \eta(X)\phi[(2n-1)\operatorname{grad}\beta - \phi(\operatorname{grad}\alpha)]$$

In a three dimensional trans-Sasakian manifolds, we have

$$\begin{aligned}Q\phi X - \phi QX &= (X\alpha)\xi - (\xi\alpha)\eta(X)\xi \\ &\quad - (\phi X)\beta\xi + \eta(X)\phi[\operatorname{grad}\beta - \phi(\operatorname{grad}\alpha)] \end{aligned}$$

Lemma 3.8 If in a three dimensional trans-Sasakian manifolds in which ξ is an eigen vector of Ricci-operator Q, then $Q\phi = \phi Q$.

References

[1] D.E Bliar, On non existence of flat contact metric strrutures, Tohoku math.journal, 28(1976), 373 − 379.
[2] Z.Olszak, On contact metric manifolds, Tohoku Math. Journ.31(1979), 247 − 253.
[3] M.Okumura, Some remarks on spaces with certain contact structures, Tohoku Math. Journ.14(1962), 135 − 145.
[4] S.Tanno, Locally symmetric K-contact Riemannian manifolds, Proc.Japan Acad.43(1967), 581-583.
[5] D.E. Blair and T. Koufogiorgos, when is the tangent sphere bundle conformally flat, J.Geom.49(1994), 55 − 66.
[6] R.Sharma, On the curvature of contact metric manifolds, J.Geom,53(1995), 179 − 190.
[7] A.Ghosh and R.Sharma, On contact strongly Pseudo-convex integrable CR manifolds, J. geom. 66(1999)116 − 122.
[8] R.S. Kulkrni, Conformally flat manifolds, Proc. Nat.Acad Sci. USA, Vol. 69,No.9 (1972), 2675 − 2676.
[9] K.Yano and S.Sawaki, Riemannian manifold admitting a conformal transformation group, J.differential Geometry 2(1968), 161 − 184.
[10] U.C. De, . and M.M. Tripati, Ricci tensor in three dimensional trans-Sasakian manifolds, Kyungpook.Math.J.2 (2005), 247 − 255.
[11] U.C. De and Shaikh A.A., Complex manifolds and contact manifolds, 2009-Narosa publication House, New Delhi.
[12] U.C. De and Shaikh A.A., Differential Geometry of manifolds, 2009-Narosa publication House, New Delhi.
[13] R. Prasad, Pankaj, M.M.Tripathi, and S.S. Shukla, On Some spetial type of trans-Sasakian manifolds, Riv. Mat.Univ. Parma(8), 2 (2009), 1 − 17.

Department of Mathematics and Astronomy
University of Lucknow, Lucknow-226007
Email address: rp.manpur@rediffmail.com,

Department of Mathematical Sciences,
Seol University, Seol-151747, South Korea
Email address: parkksn@gmail.com

Department of Mathematical and Astronomy
University of Lucknow, Lucknow-226007
Email address: drjp.oooo@gmail.com

Differential Geometry, Functional Analysis and Applications
Editors: Mohammad Hasan Shahid, Sharfuddin Ahmad *et al.*
Copyright © 2015, Narosa Publishing House, New Delhi

ON THE CAUCHY PROBLEM FOR THE NONLINEAR DIFFERENTIAL EQUATIONS WITH VALUES IN MODULAR FUNCTION SPACES

W. M. KOZLOWSKI

ABSTRACT. *We consider the Cauchy problem* $u'(t)+(I-T)u(t) = 0$, $u(0) = f$, *where an unknown function takes its values in a given modular function space, and T is a nonlinear mapping which is nonexpansive in the modular sense. It has been recently proved that under certain natural assumptions this Cauchy problem can be solved. In this paper, we demonstrate that the solution set for this problem forms a continuous nonexpansive semigroup of mappings. This result is then used to prove the existence of common fixed points of this semigroup, as well as to define some construction algorithms for such points. These results are then utilized in the construction of a stationary point for a process defined by the Cauchy problem in question. We illustrate these results by the Urysohn process extensively usedin the area of integral equations and applications.*

1. INTRODUCTION

In a recent article [51], the author proved the existence of a solution of a Cauchy problem given by a differential equations $u'(t) + (I - T)u(t) = 0$, where an unknown function takes its values in a modular function space, and T is a nonlinear mapping which is nonexpansive in the modular sense, and not necessarily in the norm sense. The purpose of the current paper is to demonstrate that the solution set for this problem forms a continuous nonexpansive semigroup of mappings.

2010 *Mathematics Subject Classification.* Primary 34G20, Secondary 34K30, 65J15, 46E30, 47H09, 46B20, 47H10, 47H20, 47J25.

Key words and phrases. Ordinary differential equation, nonlinear equation, Cauchy problem, initial value problem, fixed point, nonexpansive mapping, modular function space, semigroup of nonlinear mappings.

This result is then used to prove the existence of common fixed points of this semigroup, as well as to define some construction algorithms for such points. These results are then utilized in the construction of a stationary point for a process defined by the Cauchy problem in question. We illustrate these results by the Urysohn process which is extensively used in the area of integral equations and applications.

Modular function spaces are natural generalizations of both function and sequence variants of many important, from applications perspective, spaces like Lebesgue, Orlicz, Musielak-Orlicz, Lorentz, Orlicz-Lorentz, Calderon-Lozanovskii spaces and many others, see the book by Kozlowski [44] for an extensive list of examples and special cases.

The results and methods of fixed point theory, applied to spaces of measurable functions, have been used extensively in the field of differential and integral equations. Since the 1930s many prominent mathematicians like Orlicz and Birnbaum recognized that using the methods of L^p-spaces alone created many complications and in some cases did not allow to solve some non-power type integral equations, see [6, 62, 63]. They considered spaces of functions with some growth properties different from the power type growth control provided by the L^p-norms. Orlicz and Birnbaum considered function spaces defined as follows:

$$L^\varphi = \left\{ f : \mathbb{R} \to \mathbb{R};\ \exists \lambda > 0\ :\ \int_\mathbb{R} \varphi(\lambda|f(x)|)\ dm(x) < \infty \right\},$$

where $\varphi : [0, \infty) \to [0, \infty)$ was assumed to be a convex function increasing to infinity, i.e. the function which to some extent behaves similarly to power functions $\varphi(t) = t^p$. Let us mention two other typical examples of such functions: $\varphi_1(t) = e^t - t - 1$ or $\varphi_2(t) = e^{t^2} - 1$. The possibility of introducing the structure of a linear metric in L^φ as well as the interesting properties of these spaces, later named Orlicz spaces, and many applications to differential and integral equations with kernels of nonpower types were among the reasons for the development of the theory of Orlicz spaces, their applications and generalizations. Consider for example the following Hammerstein nonlinear integral equation which plays an important role in the elasticity theory:

$$f(x) = \int_0^1 k(x,y)\varphi(f(y))dy,$$

where $\varphi(u)$ is a function which increases more rapidly than an arbitrary power function. Krasnosel'skii and Rutickii, [55], showed that the Hammerstein operator defined by the right member of this integral equation does not operate in any of the L^p spaces. And yet, they showed how to find an Orlicz space where the Hammerstein operator is well defined and posses properties allowing to use some fixed point theorems for solving the corresponding integral equation.

Many successful applications of Orlicz spaces led to several extensions and generalizations. Using the apparatus of abstract modular spaces introduced by Nakano in [61] and then developed further by Musielak and Orlicz, see e.g. [59, 60], Musielak developed a theory of generalized Orlicz spaces, known in the contemporary literature as Musielak-Orlicz spaces, see the book by Musielak [58]. Musielak-Orlicz spaces have been proven to be very useful in application to integral and differential equations to to their flexibility and generality (e.g. they do not have to be symmetric spaces). They have been generalized to a more abstract and even more flexible setting of modular functions spaces introduced by Kozlowski in [42, 43, 44]. We refer the reader to the Section "Preliminaries" for the brief recollection of the fundamentals of the theory of modular function spaces.

There is a solid body of work on the subject of ordinary and partial differential equations in Orlicz and Musielak Orlicz spaces (but not in general modular function spaces), especially in the context of Orlicz-Sobolev and Musielak-Orlicz-Sobolev spaces, see for instance [23, 67, 24, 28, 12], and more recent works often related to applications to modeling of "smart fluids", see e.g. [66, 16, 27, 3, 25, 26, 5] and the papers referenced there. Frequently in these applications, a nonlinear extension of modular function spaces are used leading to the concepts of the modular metric spaces, see e.g. [10, 11, 1].

Typically these results use the classical techniques of differential equations with values in Banach spaces. Our approach is to use the modular notions, like the ρ-nonexpansiveness, whenever this is practical. Our results generalize the results of Khamsi from [33] obtained for norm-continuous, nonexpansive mappings acting in Musielak-Orlicz spaces L^φ with φ satisfying the Musielak-Orlicz version of the Δ_2 condition, in relation to the problem of existence of ρ-nonexpansive semigroups of mappings, which - on their own - extended classical Banach space results of [13, 65].

2. Preliminaries

Let us introduce basic notions related to modular function spaces and related notation which will be used in this paper. For further details we refer the reader to preliminary sections of the recent articles [36, 37, 14] or to the survey article [48]; see also [42, 43, 44] for the standard framework of modular function spaces.

Let Ω be a nonempty set and Σ be a nontrivial σ-algebra of subsets of Ω. Let $\mathcal{P}$ be a δ-ring of subsets of Ω, such that $E \cap A \in \mathcal{P}$ for any $E \in \mathcal{P}$ and $A \in \Sigma$. Let us assume that there exists an increasing sequence of sets $K_n \in \mathcal{P}$ such that $\Omega = \bigcup K_n$. By $\mathcal{E}$ we denote the linear space of all simple functions with supports from $\mathcal{P}$. By $\mathcal{M}_\infty$ we will denote the space of all extended measurable functions, i.e. all functions $f : \Omega \to [-\infty, \infty]$ such that there exists a sequence $\{g_n\} \subset \mathcal{E}$, $|g_n| \leq |f|$ and $g_n(\omega) \to f(\omega)$ for all $\omega \in \Omega$. By 1_A we denote the characteristic function of the set A.

Definition 2.1. *Let* $\rho : \mathcal{M}_\infty \to [0, \infty]$ *be a nontrivial, convex and even function. We say that ρ is a regular convex function pseudomodular if:*

(i) $\rho(0) = 0$;

(ii) ρ *is monotone, i.e.* $|f(\omega)| \leq |g(\omega)|$ *for all* $\omega \in \Omega$ *implies* $\rho(f) \leq \rho(g)$, *where* $f, g \in \mathcal{M}_\infty$;

(iii) ρ *is orthogonally subadditive, i.e.* $\rho(f 1_{A \cup B}) \leq \rho(f 1_A) + \rho(f 1_B)$ *for any* $A, B \in \Sigma$ *such that* $A \cap B \neq \emptyset$, $f \in \mathcal{M}$;

(iv) ρ has the Fatou property, i.e. $|f_n(\omega)| \uparrow |f(\omega)|$ for all $\omega \in \Omega$ implies $\rho(f_n) \uparrow \rho(f)$, where $f \in \mathcal{M}_\infty$;

(v) ρ is order continuous in $\mathcal{E}$, i.e. $g_n \in \mathcal{E}$ and $|g_n(\omega)| \downarrow 0$ implies $\rho(g_n) \downarrow 0$.

Similarly, as in the case of measure spaces, we say that a set $A \in \Sigma$ is ρ-null if $\rho(g1_A) = 0$ for every $g \in \mathcal{E}$. We say that a property holds ρ-almost everywhere if the exceptional set is ρ-null. As usual we identify any pair of measurable sets whose symmetric difference is ρ-null as well as any pair of measurable functions differing only on a ρ-null set. With this in mind we define $\mathcal{M} = \{f \in \mathcal{M}_\infty : |f(\omega)| < \infty\ \rho - a.e\}$, where each element is actually an equivalence class of functions equal ρ-a.e. rather than an individual function.

Definition 2.2. *We say that a regular function pseudomodular ρ is a regular convex function modular if $\rho(f) = 0$ implies $f = 0\ \rho - a.e..$ The class of all nonzero regular convex function modulars defined on Ω will be denoted by $\Re$.*

Definition 2.3. *[42, 43, 44] Let ρ be a convex function modular. A modular function space is the vector space $L_\rho = \{f \in \mathcal{M} : \rho(\lambda f) \to 0\ as\ \lambda \to 0\}$.*

The following notions will be used throughout the paper.

Definition 2.4. *Let $\rho \in \Re$.*

(a) *We say that $\{f_n\}$ is ρ-convergent to f and write $f_n \to f\ (\rho)$ if and only if $\rho(f_n - f) \to 0$.*

(b) *A sequence $\{f_n\}$ where $f_n \in L_\rho$ is called ρ-Cauchy if $\rho(f_n - f_m) \to 0$ as $n, m \to \infty$.*

(c) *A set $B \subset L_\rho$ is called ρ-closed if for any sequence of $f_n \in B$, the convergence $f_n \to f\ (\rho)$ implies that f belongs to B.*

(d) *A set $B \subset L_\rho$ is called ρ-bounded if $\sup\{\rho(f-g) : f \in B, g \in B\} < \infty$.*

(e) *A set $B \subset L_\rho$ is called strongly ρ-bounded if there exists $\beta > 1$ such that $M_\beta(B) = \sup\{\rho(\beta(f-g)) : f \in B, g \in B\} < \infty$.*

Since ρ fails in general the triangle identity, many of the known properties of limit may not extend to ρ-convergence. For example, ρ-convergence does not necessarily imply ρ-Cauchy condition. However, it is important to remember that the ρ-limit is unique when it exists. The following proposition brings together few facts that will be often used in the proofs of our results.

Proposition 2.1. *Let $\rho \in \Re$.*

(i) L_ρ is ρ-complete.

(ii) ρ-balls $B_\rho(x, r) = \{y \in L_\rho : \rho(x - y) \leq r\}$ are ρ-closed and ρ-a.e. closed.

(iii) If $\rho(\alpha f_n) \to 0$ for an $\alpha > 0$ then there exists a subsequence $\{g_n\}$ of $\{f_n\}$ such that $g_n \to 0$ ρ − a.e.

(iv) $\rho(f) \leq \liminf \rho(f_n)$ whenever $f_n \to f$ ρ−a.e. (Note: this property is equivalent to the Fatou Property).

Definition 2.5. *The following formula defines a norm in L_ρ (frequently called the* Luxemburg norm*):*

$$\|f\|_\rho = \inf\{\alpha > 0 : \rho(f/\alpha) \leq 1\}.$$

Remark 2.1. *It is not difficult to prove that $\|\cdot\|_\rho$ defines actually a norm such that $\|f\|_\rho \leq \|g\|_\rho$ whenever $|f)| \leq |g|$ ρ-a.e. It is also straightforward to demonstrate that $\|f_n\|_\rho \to 0$ if and only if $\rho(\alpha f_n) \to 0$ for every $\alpha > 0$. See Theorem 1.6 in* [58].

The above remarks immediately implies the next proposition.

Proposition 2.2. *Let $\rho \in \Re$. If $C \subset L_\rho$ is ρ-closed then C is closed with respect to the Luxemburg norm.*

Since every $\rho \in \Re$ is a left-continuous, convex modular we have the following result, see Theorems 1.5 and 1.8 in [58].

Proposition 2.3. *Let $\rho \in \Re$. The following assertions are true:*

(a) If $\|f\|_\rho < 1$ then $\rho(f) \leq \|f\|_\rho$

(b) $\|f\|_\rho \leq 1$ if and only if $\rho(f) \leq 1$.

Using the definition of the Luxemburg norm, it is easy to prove the following proposition.

Proposition 2.4. *Let $\rho \in \Re$ and let $f \in L_\rho$. Then $\|f\|_\rho > 1$ implies $\rho(f) \geq \|f\|_\rho$.*

In the sequel, we will use the following important result being an immediate corollary to Proposition 2.4.

Proposition 2.5. *Let $\rho \in \Re$. If $C \subset L_\rho$ is ρ-bounded then C is bounded with respect to the Luxemburg norm.*

We will also need the definition of the Δ_2-property of a function modular, see e.g. [44, 14].

Definition 2.6. *Let $\rho \in \Re$. We say that ρ has the Δ_2-property if*
$$\sup_n \rho(2f_n, D_k) \to 0$$
whenever $D_k \downarrow \emptyset$ and $\sup_n \rho(f_n, D_k) \to 0$.

The next result is a straightforward consequence from Definitions 2.5 and 2.6, and from Remark 2.1.

Proposition 2.6. *Let $\rho \in \Re$. The ρ-convergence is equivalent to the convergence with respect to the Luxemburg norm $\|\cdot\|_\rho$ if and only if ρ has the Δ_2-property.*

A specific subspace of L_ρ will be of particular importance for our discussions in this paper.

Definition 2.7. *Define $L_\rho^0 = \{f \in L_\rho;\ \rho(f, \cdot)\ \text{is order continuous}\}$ and $E_\rho = \{f \in L_\rho;\ \lambda f \in L_\rho^0\ \text{for every}\ \lambda > 0\}$.*

The position of E_ρ with respect to L_ρ is characterized in the following theorem.

Theorem 2.1. [42, 43, 44] *Let $\rho \in \Re$.*

(a) $L_\rho \supset L_\rho^0 \supset E_\rho$,

(b) E_ρ *has the Lebesgue property, i.e.* $\rho(\alpha f, D_k) \to 0$ *for* $\alpha > 0$, $f \in E_\rho$ *and* $D_k \downarrow \emptyset$.

(c) E_ρ is the closure of $\mathcal{E}$ (in the sense of $\|\cdot\|_\rho$).

(d) $E_\rho = L_\rho$ if and only if ρ has the Δ_2-property.

An extremal flexibility gained by using the apparatus of the modular function spaces can be illustrated as follows: the operator itself is used for the construction of a modular and hence a space in which this operator has required properties. Let us consider for instance the following Uryshon integral operator, being a generalization of the Hammerstein operator:

$$T(f)(x) = \int_0^1 k(x,y,|f(y)|)dy + f_0(x),$$

where f_0 is a fixed function and $f : [0,1] \to \mathbb{R}$ is Lebesgue measurable. For the kernel k we assume that

(a) $k : [0,1] \times [0,1] \times \mathbb{R}_+ \to \mathbb{R}_+$ is Lebesgue measurable,

(b) $k(x,y,0) = 0$,

(c) $k(x,y,.)$ is continuous, convex and increasing to $+\infty$,

(d) $\int_0^1 k(x,y,t)dx > 0$ for $t > 0$ and $y \in (0,1)$,

Assume in addition that for almost all $t \in [0,1]$ and measurable functions f, g there holds

$$\int_0^1 \left\{ \int_0^1 k(t,u,|k(u,v,|f(v)|) - k(u,v,|g(v)|)|) dv \right\} du \leq \int_0^1 k(t,u,|f(u)-g(u)|)du.$$

Setting $\rho(f) = \int_0^1 \left\{ \int_0^1 k(x,y,|f(y)|)dy \right\} dx$ and using Jensen's inequality it is easy to show that ρ is a nonnegative, even, convex nonlinear functional on the space of measurable functions $L_\rho = \{f : [0,1] \to \mathbb{R} : \exists \lambda > 0, \rho(\lambda f) < \infty\}$, and that $\rho(T(f) - T(g)) \leq \rho(f-g)$, that is, T is nonexpansive with respect to ρ. We will come back to this example towards the end of this paper, see Example 4.2.

An additional importance for applications of modular function spaces consists in the richness of structure of modular function spaces, that - besides being Banach spaces (or F-spaces in a more general settings) - are equipped with modular equivalents of norm or metric notions, and also are equipped with almost everywhere convergence and convergence in submeasure. As the above example of the Urysohn operator vividly demonstrated, in many situations in differential and integral equations, approximation and fixed point theory, modular type conditions are much more natural and modular type assumptions can be more easily verified than their metric or norm counterparts. There are also important results that can be proved only using the apparatus of modular function spaces.

Let us recall the definition of ρ-nonexpansive mappings.

Definition 2.8. *Let $\rho \in \Re$ and let $C \subset L_\rho$ be nonempty. A mapping $T : C \to C$ is called a ρ-nonexpansive mapping if $\rho(T(f) - T(g)) \leq \rho(f - g)$, for all $f, g \in C$.*

Definition 2.9. *[47] A one-parameter family $\mathcal{F} = \{T_t : t \geq 0\}$ of mappings from C into itself is said to be a ρ-nonexpansive semigroup on C if $\mathcal{F}$ satisfies the following conditions:*

(i) $T_0(x) = x$ for $x \in C$;
(ii) $T_{t+s}(x) = T_t(T_s(x))$ for $x \in C$ and $t, s \geq 0$;
(iii) for each $t \geq 0$, T_t is ρ-nonexpansive).

Definition 2.10. *A semigroup $\mathcal{F} = \{T_t : t \geq 0\}$ is called continuous if for every $z \in C$, the mapping $t \mapsto T_t(z)$ is ρ-continuous at every $t \in [0, \infty)$, i.e. $\rho\Big(T_{t_n}(z) - T_t(z)\Big) \to 0$ as $t_n \to t$.*

By $F(\mathcal{F})$ we will denote the set of common fixed points of the semigroup $\mathcal{F}$.

Let us finish this section with the existence theorems for nonexpansive mappings and semigroups of nonexpansive mappings acting in modular function spaces. First we need to recall a notion of uniformly convex function modulars.

Definition 2.11. *Let $\rho \in \Re$. We define the following uniform convexity type properties of the function modular ρ:*

(i) Let $r > 0, \varepsilon > 0$. Define

$$D_1(r, \varepsilon) = \{(f, g); f, g \in L_\rho, \rho(f) \leq r, \rho(g) \leq r, \rho(f - g) \geq \varepsilon r\}.$$

Let
$$\delta_1(r,\varepsilon) = \inf\left\{1 - \frac{1}{r}\,\rho\!\left(\frac{f+g}{2}\right); (f,g) \in D_1(r,\varepsilon)\right\}, if D_1(r,\varepsilon) \neq \emptyset,$$
and $\delta_1(r,\varepsilon) = 1$ if $D_1(r,\varepsilon) = \emptyset$.

(ii) We say that ρ is uniformly convex (UUC1) if for every $s \geq 0, \varepsilon > 0$ there exists
$$\eta_1(s,\varepsilon) > 0$$
depending on s and ε such that
$$\delta_1(r,\varepsilon) > \eta_1(s,\varepsilon) > 0 \ for \ r > s.$$

Example 2.1. *It is known that in Orlicz spaces, the Luxemburg norm is uniformly convex if and only φ is uniformly convex and Δ_2 property holds; this result can be traced to early papers by Luxemburg* [56], *Milnes* [57], *Akimovic* [2], *and Kaminska* [32]. *It is also known that, under suitable assumptions, the modular uniform convexity in Orlicz spaces is equivalent to the very convexity of the Orlicz function* [39, 9]. *Remember that the function φ is called very convex if or every $\varepsilon > 0$ and any $x_0 > 0$, there exists $\delta > 0$ such that*
$$\varphi\!\left(\frac{1}{2}(x-y)\right) \geq \frac{\varepsilon}{2}\left(\varphi(x)+\varphi(y)\right) \geq \varepsilon\varphi(x_0),$$
implies
$$\varphi\!\left(\frac{1}{2}(x+y)\right) \leq \frac{1}{2}(1-\delta)\left(\varphi(x)+\varphi(y)\right).$$

Typical examples of Orlicz functions that do not satisfy the Δ_2 condition but are very convex are: $\varphi_1(t) = e^{|t|} - |t| - 1$ and $\varphi_2(t) = e^{t^2} - 1$, [57, 55]. *Therefore, these are the examples of Orlicz spaces that are not uniformly convex in the norm sense and hence the classical Kirk theorem cannot be applied. However, these spaces are uniformly convex in the modular sense, and respective modular fixed point results can be applied.*

Theorem 2.2. [37] *Assume $\rho \in \Re$ is uniformly convex (UUC1). Let C be a ρ-closed ρ-bounded convex nonempty subset of L_ρ. Then any*

$T : C \to C$ ρ-nonexpansive mapping has a fixed point. Moreover, the set of all fixed points $F(T)$ is convex and ρ-closed.

Note that the statement of Theorem 2.2 is completely parallel to that of the Browder/Gohde/Kirk classic fixed point theorem but formulated purely in terms of function modulars without any reference to norms. Also, note that the results in [37] actually extend outside nonexpansiveness and assumes merely asymptotic pointwise ρ-nonexpansiveness of the mapping T. Therefore, Theorem 2.2 can be actually understood as the modular equivalent of the theorem by Kirk and Xu [41], see also [22, 64, 21, 71, 68, 7, 69, 70, 30, 31, 20, 40, 29, 45, 46, 49, 50, 52] and the literature referenced there.

The existence result was then extended to the existence of common fixed points of a ρ-nonexpansive semigroup of nonlinear mappings.

Theorem 2.3. [47] *Assume $\rho \in \Re$ is $(UUC1)$. Let C be a ρ-closed ρ-bounded convex nonempty subset. Let $\mathcal{F}$ be a ρ-nonexpansive semigroup on C. Then the set $F(\mathcal{F})$ of common fixed points is nonempty, ρ-closed and convex.*

3. Vitali Property

According to Proposition 2.6, the ρ-convergence and the $\|\cdot\|_\rho$-convergence are in general equivalent if only if ρ satisfies Δ_2. However, it is legitimate to ask on what subsets of L_ρ such equivalence may hold even ρ does not have Δ_2. In the paper [51] the author introduced a new concept of sets with the Vitali property that play this role. The reference to Giuseppe Vitali is justified by the following version of the Vitali Convergence Theorem which was proved in the context of modular function spaces in [44], Theorem 2.4.3.

Theorem 3.1. [44] *Let $\rho \in \Re$. Let $f_n \in E_\rho$, $f \in L_\rho$ and $f_n \to f$ ρ-a.e. Then the following conditions are equivalent:*

(i) $f \in E_\rho$ and $\|f_n - f\|_\rho \to 0$.

(ii) for every $\alpha > 0$ the subadditive measures $\rho(\alpha f_n, \cdot)$ are order equicontinuous, that is, if $E_k \in \Sigma$ are such that $E_k \downarrow \emptyset$ then
$$\lim_{k\to\infty} \sup_{n\in\mathbb{N}} \rho(\alpha f_n, E_k) = 0.$$

Definition 3.1. [51] *A set $C \subset L_\rho$ is said to posses the Vitali property if $C \subset E_\rho$, and for any $g \in L_\rho$ and $g_n \in C$ with $\rho(g_n - g) \to 0$ there exists a subsequence $\{g_{n_k}\}$ of $\{g_n\}$ such that for every $\alpha > 0$ the subadditive measures $\rho(\alpha g_{n_k}, \cdot)$ are order equicontinuous.*

Our next result characterizes sets with the Vitali property as those subsets of E_ρ on which the ρ-convergence and the $\|\cdot\|_\rho$-convergence are indeed equivalent.

Theorem 3.2. [51] *Let $\rho \in \Re$. A set $C \subset L_\rho$ has the Vitali property if and only the following two conditions are satisfied:*

(i) $C \subset E_\rho$.

(ii) If $g \in L_\rho$ and $g_n \in C$ with $\rho(g_n - g) \to 0$ then $\|g_n - g\|_\rho \to 0$.

Remark 3.1. *Combining Proposition 2.2 with Theorem 3.2, we can easily see that a set with the Vitali property is ρ-closed if and only if it is $\|\cdot\|_\rho$-closed.*

Remark 3.2. *Let $C \subset L_\rho$ be a set with the Vitali property and let $a, b \in \mathbb{R}$. Let $u : [a, b] \to C$ be a ρ-continuous function, that is, $\rho(u(t_n) - u(t)) \to 0$ provided $t_n \to t$. It follows immediately from Theorem 3.2 that u is $\|\cdot\|_\rho$-continuous.*

As an immediate corollary to Remark 3.2 we obtain the following important result.

Remark 3.3. *Let Z be a separable linear subspace of $(E_\rho, \|\cdot\|_\rho)$ and let $C \subset Z$ have the Vitali property. Assume that the function $u : [a, b] \to C$ is ρ-continuous. Then u is the Bochner integrable function with respect the the Lebesgue measure m on $[a, b]$, i.e. $u \in L^1(\Omega, Z, m)$.*

In this context, let us discuss the separability of $(E_\rho, \|\cdot\|_\rho)$. First, we need the following definition.

Definition 3.2. *The function modular $\rho \in \Re$ is called separable if $\|f\,1_{(\cdot)}\|_\rho)$ is a separable set function for each $f \in \mathcal{E}$, which means that there exists a countable $\mathcal{A} \subset \mathcal{P}$ such that to every $A \in \mathcal{P}$ there corresponds a sequence $\{A_k\}$ of elements of $\mathcal{A}$ with*

(3.1) $$\rho(\alpha f, A \Delta A_k) \to 0$$

for every $\alpha > 0$, where Δ denotes the symmetric set difference.

We are now ready to formulate the following characterization of separable $(E_\rho, \|\cdot\|_\rho)$ spaces, see Theorem 2.5.4 in [44].

Theorem 3.3. [44] *Let $\rho \in \Re$. The space $(E_\rho, \|\cdot\|_\rho)$ is a separable Banach space if and only ρ is separable.*

Finally, we can combine the last two results into the following very useful statement.

Proposition 3.1. *Let $\rho \in \Re$ be a separable function modular. If $u : [a,b] \to C$ is ρ-continuous, where $C \subset E_\rho$ has the Vitali property, then $u \in L^1(\Omega, Z, m)$.*

Let us discuss the Fatou property in the context of the Vitali property. We have the following very useful result.

Proposition 3.2. *Let $\rho \in \Re$. Assume that $C \subset E_\rho$ has the Vitali property and that*

(3.2) $$\rho(f_n - f) \to 0, \ \rho(g_n - g) \to 0$$

as $n \to \infty$, with $f_n, f, g_n, g \in C$. Then

(3.3) $$\rho(f - g) \leq \liminf_{n \to \infty} \rho(f_n - g_n).$$

Proof. Since C has the Vitali property it follows from (3.2) that

(3.4) $$\rho((f_n - g_n) - (f - g))) \to 0.$$

Using (3.4) and the Fatou property of ρ (Proposition 2.1) it is easy to get required inequality (3.3). □

Let us finish this section with few examples of sets with the Vitali property.

Example 3.1. *If ρ has Δ_2 property then every set $C \subset L_\rho$ has the Vitali property.*

Example 3.2. *Let $C \subset E_\rho$. If there exists $g \in E_\rho$ such that $|f(\omega)| \leq |g(\omega)|$ ρ-a.e for every $f \in C$ then C has the Vitali property.*

Example 3.3. *Let $C \subset E_\rho$ be $\|\cdot\|_\rho$-conditionally compact. Then C has the Vitali property (see Theorem 2.5.1 in [44]).*

4. Solution of the Initial Value Problems in Modular Function Spaces

To the end of this paper we will be considering the following initial value problem for an unknown function $u : [0, A] \to C$, where $C \subset E_\rho$:

(4.1) $$\begin{cases} u(0) = f \\ u'(t) + (I - T)u(t) = 0, \end{cases}$$

where $f \in C$ and $A > 0$ are fixed and $T : C \to C$ is ρ-nonexpansive.

Let us first consider the following question: Are the ρ-nonexpansive mappings really different from the mappings nonexpansive with respect to the Luxemburg norm associated with the modular ρ? First we will show the following simple result.

Proposition 4.1. *Let $\rho \in \Re$. If for every $\lambda > 0$*

(4.2) $$\rho\left(\lambda\left(T(f) - T(g)\right)\right) \leq \rho(\lambda(f-g))$$

then, $\|T(f) - T(g)\|_\rho \leq \|f - g\|_\rho$.

Proof. Assume to the contrary that there exist $f, g \in L_\rho$ and $\alpha > 0$ such that

$$\|f - g\|_\rho < \alpha < \|T(f) - T(g)\|_\rho.$$

Then, $\left\|\dfrac{f-g}{\alpha}\right\|_\rho < 1$, which by Proposition 2.3 part (a) implies that

$$\rho\left(\dfrac{f-g}{\alpha}\right) < 1.$$ It also implies that

$$1 < \left\|\dfrac{T(f)-T(g)}{\alpha}\right\|_\rho,$$

which, by Proposition 2.3 part (b), yields $1 < \rho\left(\dfrac{T(f)-T(g)}{\alpha}\right)$. Finally, setting $\lambda = \alpha^{-1}$, we obtain

$$\rho(\lambda(f-g)) < 1 < \rho(\lambda(T(f)-T(g))).$$

Contradiction completes the proof. □   □

In view of Proposition 4.1, we need to ask whether the inequality (4.2) needs to hold for every $\lambda > 0$ in order to ensure the norm nonexpansiveness? If we knew that it sufficed to assume it merely for $\lambda = 1$, then there would be no real reason to consider ρ-nonexpansiveness. The answer to this question can be found in the following simple example of a mapping which is ρ-nonexpansive but it is not $\|.\|_\rho$-nonexpansive.

Example 4.1. [38] *Let $X = (0,\infty)$ and Σ be the σ-algebra of all Lebesgue measurable subsets of X. Let $\mathcal{P}$ denote the δ-ring of subsets of finite measure. Define a function modular by*

$$\rho(f) = \dfrac{1}{e^2}\int_0^\infty |f(x)|^{x+1} dm(x).$$

Let B be the set of all measurable functions $f : (0,\infty) \to \mathbb{R}$ such that $0 \le f(x) \le 1/2$. Consider the map

$$T(f)(x) = \begin{cases} f(x-1), & \text{for } x \ge 1, \\ 0, & \text{for } x \in [0,1]. \end{cases}$$

Clearly, we have $T(B) \subset B$. For every $f, g \in B$ and $\lambda \le 1$, we have

$$\rho(\lambda(T(f)-T(g))) \le \lambda\rho(\lambda(f-g)),$$

which implies that T is ρ-nonexpansive. On the other hand, if we take $f = 1_{[0,1]}$, then
$$\|T(f)\|_\rho > e \geq \|f\|_\rho,$$
which clearly implies that T is not $\|.\|_\rho$-nonexpansive. Note that T is linear.

Returning to our Cauchy problem (4.1), the meaning of the above considerations is that for such mappings T the classical methods of differential equations for functions with values in Banach spaces would not work. It is also worthwhile mentioning that, due to the indirect definition of the Luxemburg norm, quite often it is much more convenient to evaluate formulas expressed only in terms of a modular which is, in applications, typically given by a direct formula allowing numerical computations.

Let us introduce the following convenient notations which will be used throughout this paper. For any $t > 0$ we define

$$(4.3) \qquad K(t) = 1 - e^{-t} = \int_0^t e^{s-t} ds.$$

We define the ρ-diameter of a set $C \subset L_\rho$ as

$$(4.4) \qquad \delta_\rho(C) = \sup_{f,g \in C} \rho(f - g).$$

Observe that $\delta_\rho(C) < \infty$ whenever the set C is ρ-bounded.

Let us start with the following technical result.

Lemma 4.1. [51] *Let $\rho \in \Re$ be separable. Let $x, y : [0, A] \to L_\rho$ be two Bochner-integrable $\| \cdot \|_\rho$-bounded functions, where $A > 0$. Then for every $t \in [0, A]$ we have*

$$(4.5) \quad \rho\left(e^{-t}y(t) + \int_0^t e^{s-t}x(s)ds\right) \leq e^{-t}\rho(y(t)) + K(t) \sup_{s \in [0,t]} \rho(x(s)).$$

In the sequel, we will be using the following key result proven in [51].

Theorem 4.1. [51] *Let $\rho \in \Re$ be separable. Let $C \subset E_\rho$ be a nonempty, convex, ρ-bounded, ρ-closed set with the Vitali property. Let $T : C \to C$ be a ρ-nonexpansive mapping. Let us fix $f \in C$ and $A > 0$ and define* the sequence of functions $u_n : [0, A] \to C$ by the following inductive formula:

(4.6) $$\begin{cases} u_0(t) = f \\ u_{n+1}(t) = e^{-t}f + \int_0^t e^{s-t}T(u_n(s))ds. \end{cases}$$

Then for every $t \in [0, A]$ there exists $u(t) \in C$ such that

(4.7) $$\rho(u_n(t) - u(t)) \to 0$$

and the function $u : [0, A] \to C$ defined by (4.7) is a solution of the Initial Value Problem (4.1). Moreover, the solution u can be extended to the whole of $[0, +\infty)$, and

(4.8) $$\rho(f - u_n(t)) \leq K^{n+1}(A)\delta_\rho(C).$$

Remark 4.1. *Theorem 4.1 extends results of [33] proven for norm continuous, nonexpansive mappings acting in Musielak-Orlicz spaces with the Δ_2 property.*

Example 4.2. *Let us go back to the example of the Urysohn operator T from the Preliminary section of this paper:*

$$T(f)(x) = \int_0^1 k(x, y, |f(y)|)dy + f_0(x),$$

As we saw, T i a ρ-nonexpansive mapping where the convex function modular ρ is defined as

(4.9) $$\rho(f) = \int_0^1 \left\{ \int_0^1 k(x, y, |f(y)|)dy \right\} dx.$$

Let us fix an $r > 0$ and set $C = \{f \in E_\rho : \rho(f - f_0) \leq r\}$. It is easy to see that $T : C \to C$. If we assume additionally that there exists a constant $M > 0$ and a Bochner-integrable function $h : [0, 1] \times [0, 1] \to [0, \infty)$ such that for every $u \geq 0$ and $x, y \in [0, 1]$

(4.10) $$k(x, y, 2u) \leq Mk(x, y, u) + h(x, y),$$

then the modular ρ has the property Δ_2 in the sense of Definition 2.6. Using the Vitali property of C and Theorem 4.1 it is then easy to see *that the corresponding Initial Value Problem*

(4.11) $$\begin{cases} u(0) = f_0 \\ u'(t) + (I - T)u(t) = 0, \end{cases}$$

has a solution in C that can be calculated as the ρ-limit of the sequence $\{u_n\}$ as defined in Theorem 4.1.

5. The Solution Set is a Continuous Semigroup

Still in the context of Theorem 4.1, we introduce the following notation: let $f \in C$, by u_f we will denote a solution of the Initial Value Problem (4.1) obtained by the ρ-limit (4.7). For any $t \geq 0$ let us define a mapping $S_t : C \to C$ by

(5.1) $$S_t(g) = u_g(t).$$

Denote $S = \{S_t\}_{t \geq 0}$ and $F(S) = \{f \in C : S_t(f) = f, \text{ for all } t \geq 0\}$. In this section we will prove that S is a continuous ρ-nonexpansive semigroup of nonlinear mappings. This result will be used in the following section for proving existence and constructing stationary points of the process defined by the system (4.1)

First let us introduce the following convenient notation:

(5.2) $$U : C \times [0, +\infty) \ni (g, t) \mapsto U(g, t) = u_g(t) = S_t(g) \in C,$$

Also, for any $n \in \mathbb{N}$ define $U_n : C \times [0, +\infty) \to C$ by following the recurrent system

(5.3) $$\begin{cases} U_0(g, t) = g \\ U_{n+1}(g, t) = e^{-t}g + \int_0^t e^{s-t} T(U_n(g, s)) ds. \end{cases}$$

First we need to prove the following important technical result.

Lemma 5.1. *Let $\rho \in \Re$ be separable. Let $C \subset E_\rho$ be a nonempty, convex, ρ-bounded, ρ-closed set with the Vitali property. Let $T : C \to C$*

be a ρ-nonexpansive mapping. Then for any $g \in C$, $t \geq 0$, $\mu \geq 0$, $n \in \mathbb{N}$, $m \in \mathbb{N}$

(5.4)
$$\rho\Big(U_n(U(g,\mu),t) - U_{n+m}(g,t+\mu)\Big) \leq \sum_{i=n+1}^{n+m+1} K^i(\mu)\delta_\rho(C) + K^{n+1}(t)\delta_\rho(C).$$

Proof. The proof is by induction on $n \in \mathbb{N}$.
Assume first that $n = 0$. Calculate

(5.5)
$$U_0(U(g,\mu),t) - U_m(g,t+\mu)$$
$$= U(g,\mu) - U_m(g,t+\mu)$$
$$= u_g(\mu) - U_m(g,t+\mu)$$
$$= u_g(\mu) - e^{-t-\mu}g - \int_0^{t+\mu} e^{s-t-\mu}T(U_{m-1}(g,s))ds.$$

Using straightforward calculus we get

(5.6)
$$\int_0^{t+\mu} e^{s-t-\mu}T(U_{m-1}(g,s))ds$$
$$= \int_0^\mu e^{s-\mu}T(U_{m-1}(g,s))ds + \int_0^t e^s T(U_{m-1}(g,s+\mu))ds.$$

Using the definition of U_m and the formula (5.6) we obtain the following

(5.7) $\quad U_m(g,t+\mu) = e^{-t}\Big(e^{-\mu} + \int_0^\mu e^{s-\mu}T(U_{m-1}(g,s))ds\Big)$
$$+ e^{-t}\int_0^t e^s T(U_{m-1}(g,s+\mu))ds$$
$$= e^{-t}U_m(g,\mu) + e^{-t}\int_0^t e^s T(U_{m-1}(g,s+\mu))ds.$$

Substituting (5.7) into (5.5) we get

$$(5.8) \quad U_0(U(g,\mu),t) - U_m(g,t+\mu)$$

$$= u_g(\mu) - e^{-t}U_m(g,\mu) - e^{-t}\int_0^t e^s T(U_{m-1}(g,s+\mu))ds$$

$$= e^{-t}\Big(u_g(\mu) - U_m(g,\mu)\Big) + \int_0^t e^{s-t}\Big(u_g(\mu) - T(U_{m-1}(g,s+\mu))\Big)ds,$$

where we used the fact that

$$(5.9) \quad e^{-t}u_g(\mu) + \int_0^t e^{s-t}u_g(\mu)ds = u_g(\mu).$$

Let us apply Lemma 4.1 with

$$x(t) = u_g(\mu) - T(U_{m-1}(g,t+\mu)),$$
$$y(t) = u_g(\mu) - U_m(g,\mu).$$

Hence, using (4.5) and (5.8) we get

$$(5.10) \quad \rho\Big(U_0(U(g,\mu),t) - U_m(g,t+\mu)\Big) = \rho\Big(e^{-t}y(t) + \int_0^t e^{s-t}x(s)ds\Big)$$

$$\leq e^{-t}\rho(y(t)) + K(t)\sup_{0\leq s\leq t}\rho(x(s))$$

$$= e^{-t}\rho(u_g(\mu) - U_m(g,\mu)) + K(t)\sup_{0\leq s\leq t}\rho\Big(u_g(\mu) - T(U_{m-1}(g,s+\mu))\Big)$$

$$\leq e^{-t}\rho(u_g(\mu) - U_m(g,\mu)) + K(t)\delta_\rho(C).$$

Applying to (5.10) inequality (4.8) we finally arrive at

$$(5.11) \quad \rho\Big(U_0(U(g,\mu),t) - U_m(g,t+\mu)\Big) \leq K^{m+1}(t)\delta_\rho(C) + K(t)\delta_\rho(C),$$

which gives us desired inequality (5.4) with $n = 0$.

Assume now that (5.4) is true for $n \in \mathbb{N}$ and let us prove it for $n+1$. Using the definition of the recurrent sequence combined with (5.7) we have

(5.12) $$U_{n+1}(U(g,\mu),t) - U_{n+m+1}(g,t+\mu)$$

$$= e^{-t}U(g,\mu) + \int_0^t e^{s-t}T(U_n(U(g,\mu),s))ds - U_{n+m+1}(g,t+\mu)$$

$$= e^{-t}U(g,\mu) + \int_0^t e^{s-t}T(U_n(U(g,\mu),s))ds$$

$$- \left(e^{-t}U_{n+m+1}(g,\mu) + e^{-t}\int_0^t e^s T(U_{n+m}(g,s+\mu))ds\right)$$

$$= e^{-t}\Big(U(g,\mu) - U_{n+m+1}(g,\mu)\Big) + \int_0^t e^{s-t}\Big(T(U_n(U(g,\mu),s)) - T(U_{n+m}(g,s+\mu))\Big)ds.$$

Applying Lemma 4.1 with

$$x(t) = T(U_n(U(g,\mu),t)) - T(U_{n+m}(g,t+\mu)),$$
$$y(t) = U(g,\mu) - U_{n+m+1}(g,\mu),$$

we conclude from (5.12) that

(5.13) $$\rho\Big(U_{n+1}(U(g,\mu),t) - U_{n+m+1}(g,t+\mu)\Big)$$

$$\leq e^{-t}\rho(U(g,\mu) - U_{n+m+1}(g,\mu))$$

$$+ K(t) \sup_{0 \leq s \leq t} \rho\Big(T(U_n(U(g,\mu),s)) - T(U_{n+m}(g,s+\mu))\Big).$$

Using the ρ-nonexpansiveness of T, the inductive assumption and (4.8) we conclude from (5.13) that

(5.14) $$\rho\Big(U_{n+1}(U(g,\mu),t) - U_{n+m+1}(g,t+\mu)\Big)$$

$$\leq e^{-t}\rho(U(g,\mu) - U_{n+m+1}(g,\mu)) + K(t)\Big(\sum_{i=m+1}^{n+m+1} K^i(\mu) + K^{n+1}(t)\Big)\delta_\rho(C)$$

$$\leq e^{-t}K^{n+m+2}(\mu)\delta_\rho(C) + K(t)\Big(\sum_{i=m+1}^{n+m+1} K^i(\mu) + K^{n+1}(t)\Big)\delta_\rho(C)$$

(5.15) $$\leq \sum_{i=m+1}^{n+m+2} K^i(\mu)\delta_\rho(C) + K^{n+2}(t)\delta_\rho(C),$$

which completes the proof of Lemma 5.1. □

We are now ready to prove the main result of this section:

Theorem 5.1. *Let $\rho \in \Re$ be separable, $C \subset E_\rho$ be a nonempty, convex, ρ-bounded, ρ-closed set with the Vitali property. Assume $T : C \to C$ to be a ρ-nonexpansive mapping. Denote $S_t(f) = u_f(t)$, where $t \geq 0$, $f \in C$ and $u_f(t)$ is a solution of the Initial Value Problem (4.6). Then $\{S_t\}_{t \geq 0}$ is a continuous ρ-nonexpansive semigroup of non-linear mappings on C.*

Proof. First let us fix $t \geq 0$ and demonstrate that S_t is ρ-nonexpansive. Fix $f, g \in C$ and recall that by Theorem 4.1,

(5.16)
$$\rho(U_n(f,t) - U(f,t)) \to 0$$
$$\rho(U_n(g,t) - U(g,t)) \to 0$$

as $n \to \infty$. By Proposition 3.2 it follows from (5.16) that

(5.17) $\rho(U(f,t) - U(g,t)) \leq \liminf_{n \to \infty} \rho(U_n(f,t) - U_n(g,t)).$

Applying Lemma 4.1 with

$$x(t) = U_n(f,t) - U_n(g,t),$$
$$y(t) = f - g,$$

we have

(5.18)
$$\rho(U_{n+1}(f,t) - U_{n+1}(g,t))$$
$$= \rho\left(e^{-t}f + \int_0^t e^{s-t}U_n(f,s)ds - e^{-t}g - \int_0^t e^{s-t}U_n(g,s)ds\right)$$
$$\leq e^{-t}\rho(f-g) + K(t) \sup_{0 \leq s \leq t} \rho(U_n(f,s) - U_n(g,s)).$$

By obvious induction, keeping in mind that $K(t) = 1 - e^{-t}$ we can easily deduce from (5.18) that

(5.19) $\rho(U_n(f,t) - U_n(g,t)) \leq \rho(f-g),$

which together with (5.17) gives us finally

(5.20) $$\rho(S_t(f) - S_t(g)) = \rho(U(f,t) - U(g,t)) \leq \rho(f - g),$$

hence S_t is ρ-nonexpansive.

Now let us prove that $\{S_t\}_t \geq 0$ forms a semigroup of non-linear mappings. Since it is obvious that $S_0(f) = u_f(0) = f$, it remain to prove that

(5.21) $$S_{\mu+t} = S_\mu \circ S_t$$

for any $\mu, t \geq 0$. Let us fix temporarily $n \in \mathbb{N}$ and note that by Theorem 4.1

(5.22) $$\rho(U_{n+m}(f, t + \mu) - U(f, t + \mu)) \to 0$$

as $m \to \infty$. Hence, by Proposition 3.2 and by inequality (5.4) from Lemma 5.1, using also the fact that the series $\sum_i K^i(\mu)$ is convergent, we obtain the following

(5.23) $$\rho\Big(U_n(U(f,\mu),t) - U(f, t+\mu)\Big)$$
$$\leq \liminf_{m \to \infty} \rho\Big(U_n(U(f,\mu),t) - U_{n+m}(f, t+\mu)\Big)$$
$$\leq \liminf_{m \to \infty} \sum_{i=n+1}^{n+m+1} K^i(\mu)\delta_\rho(C) + K^{n+1}(t)\delta_\rho(C)$$
$$\leq K^{n+1}(t)\delta_\rho(C).$$

From (5.23) we see that

(5.24) $$\rho\Big(U_n(U(f,\mu),t) - S_{t+\mu}(f)\Big) = \rho\Big(U_n(U(f,\mu),t) - U(f,t+\mu)\Big) \to 0$$

as $n \to \infty$. On the other hand, it follows from Theorem 4.1 that

(5.25) $$\rho\Big(U_n(U(f,\mu),t) - S_t(S_\mu(f))\Big) = \rho\Big(U_n(S_\mu(f),t) - S_t(S_\mu(f))\Big) \to 0$$

as well. The uniqueness of the ρ-limit yields to $S_{t+\mu}(f) = S_t(S_\mu(f))$ for every $t, \mu \geq 0$ and every $f \in C$. To prove ρ-continuity of the semigroup

$\{S_t\}_{t\geq 0}$ let us observe that if $t_n \to t$ then

(5.26) $\qquad \rho(S_{t_n}(f) - S_t(f)) = \rho(u_f(t_n) - u_f(t)) \to 0$

because u_f is continuous as a solution of the differential equation (4.6). This completes the proof of Theorem 5.1. $\qquad \square$

6. Application to Existence and Construction of Stationary Points

In the previous section we proved that $\mathcal{S} = \{S_t\}_{t\geq 0}$, where $S_t(g) = u_g(t)$ for any $g \in C$, is a continuous ρ-nonexpansive semigroup of nonlinear mappings. Such a situation is quite typical in mathematics and applications. For instance, in the theory of dynamical systems, the modular function space L_ρ would define the state space and the mapping $(t,x) \to S_t(x)$ would represent the evolution function of a dynamical system. The question about the existence of common fixed points, and about the structure of the set of common fixed points, can be interpreted as a question whether there exist stationary points for this process, that is, elements of C that are fixed during the state space transformation S_t at any given point of time t, and if yes - what the structure of a set of such points may look like and how such points can be constructed algorithmically. In the setting of this paper, the state space may be an infinite dimensional. Therefore, it is natural to apply these result to not only to deterministic dynamical systems but also to stochastic dynamical systems.

It is immediate that if $F \in C$ is a common fixed point of $\mathcal{S}$ then $S_t(f) = f$ for all $t \geq 0$ which means that $u_f(t) = f$ for every such t. Therefore, the process defined by the equation (4.6) has a stationary point at such f. Let us now interpret our previous results in this context.

Combining Theorem 2.3, Theorem 4.1 and Theorem 5.1, we arrive at the following result.

Theorem 6.1. *Let $\rho \in \Re$ be separable and $(UUC1)$, $C \subset E_\rho$ be a nonempty, convex, ρ-bounded, ρ-closed set with the Vitali property. Assume $T : C \to C$ to be a ρ-nonexpansive mapping. Denote $S_t(f) =*

$u_f(t)$, where $t \geq 0$, $f \in C$ and $u_f(t)$ is a solution of the Initial Value Problem (4.6). Then $\mathcal{S} = \{S_t\}_{t \geq 0}$ forms a continuous ρ-nonexpansive semigroup of non-linear mappings on C. Moreover, the set of the stationary points for the process defined by the differential equation (4.6) with the evolution function $(t, x) \to S_t(x)$, is equal to $F(\mathcal{S})$, the set of all common fixed points of $\mathcal{S}$, which is nonempty, ρ-closed and convex.

Proving existence is one thing but being able to actually construct algorithmically such a stationary point is a completely different proposition. Fortunately in some cases this daunting task can be replaced by a simpler task of constructing a fixed point for just one ρ-nonexpansive mapping. To this end let us quote the following very recent representation result.

Theorem 6.2. [4] Let $\rho \in \Re$ be (UUC), and let $\mathcal{F} = \{T_t : t \geq 0\}$ be a continuous semigroup of ρ-nonexpansive mappings on a ρ-closed, ρ-bounded, convex, nonempty subset of L_ρ. Let $\alpha > 0$ and $\beta > 0$ be two real numbers such that $\alpha/\beta \notin \mathbb{Q}$. Fix an arbitrary $\lambda \in (0, 1)$. Then

(6.1) $$F(\mathcal{F}) = F\Big(\lambda T_\alpha + (1 - \lambda) T_\beta\Big).$$

Remark 6.1. Using different methods, the conclusion of Theorem 6.2 can be proved without the assumption of the uniform convexity but assuming instead the strong continuity of the semigroup $\mathcal{F}$, see Theorem 3.1 in [54].

In view of Theorem 6.2 and Remeark 6.1, in order to construct a stationary point for a process defined by (4.6) it is enough (under some reasonable assumptions) to construct a fixed point for just one ρ-nonexpansive mapping. There are several known algorithms, based on the Mann and Ishikawa processes, that can be used for such construction, see e.g. [14, 15, 4].

The above-mentioned results establish a connection between the theory of differential equations in modular function spaces and the fixed point theory for nonlinear mappings acting in modular function spaces. The latter theory has been a subject to intensive study since 1990s, see e.g. [38, 39, 34, 35, 17, 18, 19, 36, 37, 47, 48, 14, 54, 53, 51, 15, 4].

Let us finish this section and the paper by providing an example how the results of the preceding sections can be utilized for constructing a stationary point of a process defined by the Urysohn operator:

$$T(f)(x) = \int_0^1 k(x, y, |f(y)|)dy + f_0(x),$$

with the assumptions as per Example 4.2. Using Theorem 4.1 we see that, given $f \in C$, the Initial Value Problem

(6.2) $$\begin{cases} u(0) = f \\ u'(t) + (I - T)u(t) = 0, \end{cases}$$

has a solution $u_f : [0, +\infty] \to C$. As proved in our Theorem 5.1 the formula

$$S_t(f) = u_f(t)$$

defines the semigroup of ρ-nonexpansive mappings. Note that ρ in this example is orthogonally additive and hence it has the Strong Opial Property, see [35]. Therefore, assuming ρ is $(UUC1)$ and uniformly continuous, see [58, 38, 9] for several criteria, we can use the algorithmic methods from [4] (generalized Mann process in Theorem 4.3, and modified Ishikawa process in Theorem 5.2) to construct a common fixed point of the semigroup $\{S_t\}$ which will be a stationary point of the Urysohn process defined by the evolution function $(t, f) \to u_f(t) \in C$.

REFERENCES

[1] A.A.N. Abdou, and M.A. Khamsi, *On the fixed points of nonexpansive maps in modular metric spaces*, Fixed Point Theory and Applications 2013, 2013:229.

[2] B. A. Akimovic, *On uniformly convex and uniformly smooth Orlicz spaces*, Teor. Funkc. Funkcional. Anal. i Prilozen., 15 (1972).

[3] M.K. Alaoui, *On Elliptic Equations in Orlicz Spaces Involving Natural Growth Term and Measure Data*, Abstract and Applied Analysis, 2012:615816 (2012).

[4] M. Alsulami, and W.M. Kozlowski *On the set of common fixed points of semigroups of nonlinear mappings in modular function spaces*, Fixed Point Theory and Applications 2014, 2014:4.

[5] A. Benkirane and M. Sidi El Vally, *An existence result for nonlinear elliptic equations in Musielak-Orlicz-Sobolev spaces*, Bull. Belg. Math. Soc. Simon Stevin, 20.1 (2013), 57 - 75.

[6] Z. Birnbaum, and W. Orlicz, *Uber die Verallgemeinerung des Begriffes der zueinander konjugierten Potenzen*, Studia Math., 3 (1931), 1 - 67.

[7] R.Bruck, T. Kuczumow, and S. Reich, *Convergence of iterates of asymptotically nonexpansive mappings in Banach spaces with the uniform Opial property*, Coll. Math., 65.2 (1993), 169 - 179.

[8] J. Cerda, H. Hudzik, and M. Mastylo, *On the geometry of some Calderon-Lozanovskii interpolation spaces*, Indagationes Math., 6.1 (1995), 35 - 49.

[9] S. Chen, *Geometry of Orlicz Spaces*, Dissertationes Mathematicae, 356 (1996).

[10] V.V. Chistyakov, *Modular metric spaces, I: Basic concepts*, Nonlinear Analysis, 72.1 (2010), 1-14.

[11] V.V. Chistyakov, *Modular metric spaces, II: Application to superposition operators*, Nonlinear Analysis, 72.1 (2010), 15-30.

[12] A. Cianchi, *Optimal Orlicz-Sobolev embeddings*, Revista Mathematica Iberoamericana, 20.2 (2004), 427 - 474.

[13] M.C. Crandall, and A. Pazy, *Semigroups of nonlinear contractions and dissipative sets*, J. Funct. Anal., 3 (1963), 376 - 418.

[14] B. A. Bin Dehaish, and W.M. Kozlowski, *Fixed point iterations processes for asymptotic pointwise nonexpansive mappings in modular function spaces*, Fixed Point Theory and Applications, 2012:118 (2012).

[15] B. A. Bin Dehaish, M.A. Khamsi, and W.M. Kozlowski, *Common fixed points for asymptotic pointwise Lipschitzian semigroups in modular function spaces*, Fixed Point Theory and Applications, 2013:214 (2013).

[16] L. Diening, *Theoretical and numerical results for electrorheological fluids*, Ph. D. Thesis (2002), University of Freiburg, Germany.

[17] T. Dominguez-Benavides, M.A. Khamsi, and S. Samadi, *Uniformly Lipschitzian mappings in modular function spaces*, Nonlinear Analysis, 46 (2001), 267-278.

[18] T. Dominguez-Benavides, M.A. Khamsi, and S. Samadi, *Asymptotically regular mappings in modular function spaces*, Scientiae Mathematicae Japonicae, 53 (2001), 295-304.

[19] T. Dominguez-Benavides, M.A. Khamsi, and S. Samadi, *Asymptotically nonexpansive mappings in modular function spaces*, J. Math. Anal. Appl., 265.2 (2002), 249-263.

[20] J. Garcia Falset, W. Kaczor, T. Kuczumow and S. Reich, *Weak convergence theorems for asymptotically nonexpansive mappings and semigroups*, Nonlinear Analysis, 43 (2001), 377-401.

[21] J. Gornicki, *Weak convergence theorems for asymptotically nonexpansive mappings in uniformly convex Banach spaces*, Comment. Math. Univ. Carolin., 30 (1989), 249 - 252.

[22] K. Goebel, and W.A. Kirk, *A fixed points theorem for asymptotically nonexpansive mappings*, Proc. Amer. Math. Soc., 35.1 (1972), 171 - 174.

[23] J.-P. Gossez, *Nonlinear elliptic boundary value problems for equations with rapidly (or slowly) increasing coefficients*, Trans. Amer. Math. Soc., 190 (1974), 163 - 205.

[24] J.-P. Gossez, *Some approximation properties in Orlicz-Sobolev spaces*, Studia Math., 74.1 (1982), 17-24.

[25] P. Gwiazda, P. Minakowski, and A. Wroblewska-Kaminska, *Elliptic problems in generalized Orlicz-Musielak spaces*, CEJM 2012.

[26] P. Gwiazda, P. Wittbold, A. Wroblewska, and A. Zimmermann, *Renormalized solutions of nonlinear elliptic problems in generalized Orlicz spaces*, Journal of Differential Equations, 253 (2012), 635-666.

[27] P. Harjulehto, P. Hasto, M. Koskenoja, and S. Varonen *The Dirichlet Energy Integral and Variable Exponent Sobolev Spaces with Zero Boundary Values*, Potential Analysis, 25.3 (2006), 205-222.

[28] J. Heinonen, T. Kilpelainen, and O. Martio, *Nonlinear potential theory of degenerate elliptic equations*, Oxford University Press (1993), Oxford.

[29] N. Hussain, and M.A. Khamsi, *On asymptotic pointwise contractions in metric spaces*, Nonlinear Analysis, 71.10 (2009), 4423 - 4429.

[30] W. Kaczor, T. Kuczumow and S. Reich, *A mean ergodic theorem for nonlinear semigroups which are asymptotically nonexpansive in the intermediate sense*, J. Math. Anal. Appl., 246 (2000), 1 - 27.

[31] W. Kaczor, T. Kuczumow and S. Reich, *A mean ergodic theorem for mappings which are asymptotically nonexpansive in the intermediate sense*, Nonlinear Analysis, 47 (2001), 2731-2742.

[32] A. Kaminska, *On uniform convexity of Orlicz spaces*, Indag. Math. 44.1 (1982), 27-36.

[33] M.A. Khamsi, *Nonlinear semigroups in modular function spaces*, Math. Japonica, 37.2 (1992), 1-9.

[34] M.A. Khamsi, *Fixed point theory in modular function spaces*, Proceedings of the Workshop on Recent Advances on Metric Fixed Point Theory held in Sevilla, September, 1995, 31-35. MR1440218(97m:46044).

[35] M.A. Khamsi, *A convexity property in modular function spaces*, Math. Japonica, 44.2 (1996), 269-279.

[36] M.A. Khamsi, and W.M. Kozlowski, *On asymptotic pointwise contractions in modular function spaces*, Nonlinear Analysis, 73 (2010), 2957 - 2967.

[37] M.A. Khamsi, and W.M. Kozlowski, *On asymptotic pointwise nonexpansive mappings in modular function spaces*, J. Math. Anal. Appl., 380.2 (2011), 697 - 708.

[38] M.A. Khamsi, W.M. Kozlowski, and S. Reich, *Fixed point theory in modular function spaces*, Nonlinear Analysis, 14 (1990), 935-953.

[39] M.A. Khamsi, W.M. Kozlowski, and S. Chen, *Some geometrical properties and fixed point theorems in Orlicz spaces*, J. Math. Anal. Appl., 155.2 (1991), 393-412.

[40] W.A. Kirk, *Asymptotic pointwise contractions*, in: Plenary Lecture, the 8th International Conference on Fixed Point Theory and Its Applications, Chiang Mai University, Thailand, July 16-22, 2007.

[41] W. A. Kirk, and H.K. Xu, *Asymptotic pointwise contractions*, Nonlinear Anal., 69 (2008), 4706 - 4712.

[42] W.M. Kozlowski, *Notes on modular function spaces I*, Comment. Math., 28 (1988), 91-104.

[43] W.M. Kozlowski, *Notes on modular function spaces II*, Comment. Math., 28 (1988), 105-120.

[44] W.M. Kozlowski, *Modular Function Spaces*, Series of Monographs and Textbooks in Pure and Applied Mathematics, Vol.122, Dekker, New York/Basel, 1988.

[45] W.M. Kozlowski, *Fixed point iteration processes for asymptotic pointwise nonexpansive mappings in Banach spaces*, J. Math. Anal. Appl., 377.1 (2011), 43 - 52.

[46] W.M. Kozlowski, *Common fixed points for semigroups of pointwise Lipschitzian mappings in Banach spaces*, Bull. Austral. Math Soc., 84 (2011), 353 - 361.

[47] W.M. Kozlowski, *On the existence of common fixed points for semigroups of nonlinear mappings in modular function spaces*, Comment. Math., 51.1 (2011), 81 - 98.

[48] W.M. Kozlowski, *Advancements in fixed point theory in modular function*, Arab J. Math., (2012), doi:10.1007/s40065-012-0051-0.

[49] W.M. Kozlowski, *On the construction of common fixed points for semigroups of nonlinear mappings in uniformly convex and uniformly smooth Banach spaces*, Comment. Math., 52.2 (2012), 113 - 136.

[50] W.M. Kozlowski, *Pointwise Lipschitzian mappings in uniformly convex and uniformly smooth Banach spaces*, Nonlinear Analysis, 84 (2013), 50 - 60.

[51] W.M. Kozlowski *On Nonlinear Differential Equations in Generalized Musielak-Orlicz Spaces*, Comment. Math., 53.2 (2013), 113 - 133.

[52] W.M. Kozlowski, and B. Sims *On the convergence of iteration processes for semigroups of nonlinear mappings in Banach spaces*, In: Bailey, DH, Bauschke, HH, Borwein, P, Garvan, F, Thera, M, Vanderwerff, JD, Wolkowicz, H (eds.) Computational and Analytical Mathematics. In Honor of Jonathan Borweins 60th Birthday. Springer Proceedings in Mathematics and Statistics, vol. 50. Springer, New York (2013).

[53] W.M. Kozlowski *An Introduction to Fixed Point Theory in Modular Function Spaces*, In "Topics in Fixed Point Theory", Ed: S. Almezel, Q.H. Ansari, M.A. Khamsi, Springer Verlag, New York Heidelberg Dordrecht London, 2014.

[54] W.M. Kozlowski *On common fixed points of semigroups of mappings nonexpansive with respect to convex function modulars*, J. Nonlinear Convex Anal., 15.4(2014), 437-449.

[55] M.A. Krasnosel'skii, and Y.B. Rutickii, *Convex Functions and Orlicz Spaces.* P. Nordhoff Ltd, Groningen, 1961.

[56] W.A.J. Luxemburg, *Banach Function Spaces.* Thesis, Delft, (1955).

[57] H.W. Milnes, *Convexity of Orlicz spaces. Pacific J. Math.*, 7 (1957), 1451-1486.

[58] J. Musielak, *Orlicz Spaces and Modular Spaces.* Lecture Notes in Mathematics, Vol. 1034, Springer-Verlag, Berlin/Heidelberg/New York/Tokyo, 1983.

[59] J. Musielak, and W. Orlicz, *On modular spaces.* Studia Math., 18, 49 – 65 (1959).

[60] J. Musielak, and W. Orlicz, *Some remarks on modular spaces.* Bull. Acad. Polon. Sci. Ser. Sci. Math. Astronom. Phys., 7 (1959), 661 - 668.

[61] Nakano, H.: *Modulared Semi-ordered Linear Spaces.* Maruzen Co., Tokyo, 1950.

[62] W. Orlicz, *Uber eine gewisee klasse von Raumen vom Typus B,* Bull. Acad. Polon. Sci. Ser. A, (1932), 207 - 220.

[63] W. Orlicz, *Uber Raumen L^M,* Bull. Acad. Polon. Sci. Ser. A, (1936), 93 - 107.

[64] S. Reich, *Weak convergence theorems for nonexpansive mappings in Banach spaces,* J. Math. Anal. Appl., 67 (1979), 274 - 276.

[65] S. Reich, *A note on the mean ergodic theorem for nonlinear semigroups,* J. Math. Anal. Appl., 91 (1983), 547 - 551.

[66] M. Ruzicka, *Electrorheological Fluids: Modeling and Mathematical Theory,* Lecture Notes in Mathematics 1748 (2000), Springer Verlag, Berlin.

[67] G. Talenti, *Nonlinear elliptic equations, rearrangements of functions and Orlicz spaces,* Annali di Matematica Pura ed Applicata, 120 (1979), 159 - 184.

[68] K-K.Tan, and H-K. Xu, *An ergodic theorem for nonlinear semigroups of Lipschitzian mappings in Banach spaces,* Nonlinear Anal., 19.9 (1992), 805 - 813.

[69] K-K.Tan, and H-K. Xu, *Approximating fixed points of nonexpansive mappings by the Ishikawa iteration process,* J. Math. Anal. Appl., 178 (1993), 301 - 308.

[70] K-K.Tan, and H-K. Xu, *Fixed point iteration processes for asymptotically nonexpansive mappings,* Proc. Amer. Math. Soc., 122 (1994), 733 - 739.

[71] H-K. Xu, *Existence and convergence for fixed points of asymptotically nonexpansive type,* Nonlinear Anal., 16 (1991), 1139 - 1146.

W. M. KOZLOWSKI, SCHOOL OF MATHEMATICS AND STATISTICS, UNIVERSITY OF NEW SOUTH WALES, SYDNEY, NSW 2052, AUSTRALIA

E-mail address: w.m.kozlowski@unsw.edu.au

Differential Geometry, Functional Analysis and Applications
Editors: Mohammad Hasan Shahid, Sharfuddin Ahmad *et al.*
Copyright © 2015, Narosa Publishing House, New Delhi

COINCIDENCES AND COMMON FIXED POINT THEOREMS IN INTUITIONISTIC FUZZY METRIC SPACES USING GENERAL CONTRACTIVE CONDITION OF INTEGRAL TYPE

AMIT SINGH AND B. FISHER

ABSTRACT. In the present paper, we first of all prove a coincidence theorem for a family of mappings on an arbitrary set with values in an intuitionistic fuzzy metric space. We further establish a common fixed point theorem. Our results generalize and extend some of the well known results in metric and other spaces.

1. INTRODUCTION

Atanassov [4] introduced and studied the concept of intuitionistic fuzzy sets as a generalization of fuzzy sets. Since then, there has been much progress in the study of intuitionistic fuzzy sets by many authors, see for instance [5], [6], [7] and [8]. Recently, J.H. Park [13] has introduced and studied the notion of intuitionistic fuzzy metric spaces. In [3] Alaca, Turkoglu and Yildiz, they proved the well known fixed point theorems of Banach and Edelstein in intuitionistic fuzzy metric spaces with the help of Grabiec [9]. Further, Jesic and Babacev [10] proved some common fixed point theorems for a pair of R-weakly commuting mappings on this newly defined space.

The present paper is concerned with extending and generalizing the theory of fixed points to intuitionistic fuzzy metric spaces. We first of all prove a coincidence theorem for a family of mappings on an arbitrary set with values in an intuitionistic fuzzy metric space. We further establish a common fixed point theorem for a family of self mappings along with the mappings P, Q, S and T. We generalize and extend some of the results of Jesic & Babacev

2000 *Mathematics Subject Classification.* 47H10, 54H25.
Key words and phrases. Intuitionistic fuzzy metric spaces, Coincidence points, Common fixed point.

[10] and Mishra, Singh & Chadha [12] to intuitionistic fuzzy metric spaces. Our results fuzzify and generalize several results on metric, fuzzy and Menger spaces.

2. PRELIMINARIES

Definition 2.1. [14]. *A binary operation* $*: [0,1] \times [0,1] \to [0,1]$ *is a continuous t norm if* $*$ *satisfies the following conditions:*
(a) $*$ *is commutative and associative,*
(b) $*$ *is continuous,*
(c) $a * 1 = a$ *for all* $a \in [0,1]$,
(d) $a * b \leq c * d$ *whenever* $a \leq c$ *and* $b \leq d$ *and* $a, b, c, d \in [0,1]$.

Definition 2.2. [14]. *A binary operation* $\diamond: [0,1] \times [0,1] \to [0,1]$ *is a continuous t-conorm if* $\diamond$ *satisfies the following conditions:*
(a) $\diamond$ *is commutative and associative,*
(b) $\diamond$ *is continuous,*
(c) $a \diamond 0 = a$ *for all* $a \in [0,1]$,
(d) $a \diamond b \leq c \diamond d$ *whenever* $a \leq c$ *and* $b \leq d$ *and* $a, b, c, d \in [0,1]$.

Definition 2.3. [3]. *A 5-tuple* $(X, M, N, *, \diamond)$ *is said to be an intuitionistic fuzzy metric space if* X *is an arbitrary set,* $*$ *is a continuous t-norm,* $\diamond$ *is a continuous t-conorm and* M N *are fuzzy sets on* $X^2 \times [0, \infty)$ *satisfying the following conditions:*
(i) $M(x, y, t) + N(x, y, t) \leq 1$,
(ii) $M(x, y, 0) = 0$,
(iii) $M(x, y, t) = 1$ *for all* $t > 0$ *iff* $x = y$,
(iv) $M(x, y, t) = M(y, x, t)$,
(v) $M(x, y, t) * M(y, z, s) \leq M(x, z, t + s)$ *for all* $x, y, z \in X$, $s, t > 0$,
(vi) $M(x, y, \bullet) : [0, \infty) \to [0,1]$ *is left continuous,*
(vii) $\lim_{t \to \infty} M(x, y, t) = 1$ *for all* x, y in X,
(viii) $N(x, y, 0) = 1$,
(ix) $N(x, y, t) = 0$ *for all* $t > 0$ *iff* $x = y$,
(x) $N(x, y, t) = N(y, x, t)$,
(xi) $N(x, y, t) \diamond N(y, z, s) \geq N(x, z, t + s)$ *for all* $x, y, z \in X$ *and* $s, t > 0$,
(xii) $N(x, y, \bullet) : [0, \infty) \to [0,1]$ *is right continuous,*
(xiii) $\lim_{t \to \infty} N(x, y, t) = 0$ *for all* x, y *in* X.
Then (M, N) *is called an intuitionistic fuzzy metric on* X. *The*

functions $M(x,y,t)$ and $N(x,y,t)$ denote the degree of nearness and degree of non-nearness between x and y with respect to t, respectively.

In an intuitionistic fuzzy metric space X, $M(x,y,\bullet)$ is non-decreasing and $N(x,y,\bullet)$ is non-increasing for all x,y in X.

Example 2.1. [13]. *Let (X,d) be a metric space. For every $a,b \in [0,1]$ with $a*b = a.b$ and $a \diamond b = \min\{1, a+b\}$, let $M(x,y,t) = \frac{t}{t+d(x,y)}$ and $N(x,y,t) = \frac{d(x,y)}{t+d(x,y)}$ for all $x,y \in X$ and $t > 0$. Then $(X, M, N, *, \diamond)$ is an intuitionistic fuzzy metric space induced by the metric d. It is obvious that $N(x,y,t) = 1 - M(x,y,t)$.*

Definition 2.4. [3]. *Let $(X, M, N, *, \diamond)$ be an intuitionistic fuzzy metric space. Then*

(A) a sequence $\{x_n\}$ in X is said to be Cauchy sequence if for each $t > 0$ and $p > 0$

$$\lim_{n \to \infty} M(x_{n+p}, x_n, t) = 1 \text{ and } \lim_{n \to \infty} N(x_{n+p}, x_n, t) = 0,$$

(B) a sequence $\{x_n\}$ in X is converging to x in X if for each $t > 0$,

$$\lim_{n \to \infty} M(x_n, x, t) = 1 \text{ and } \lim_{n \to \infty} N(x_n, x, t) = 0.$$

An intuitionistic fuzzy metric space is said to be complete if and only if every Cauchy sequence is convergent.

Lemma 2.1. [11]. *Let $(X, M, N, *, \diamond)$ be an intuitionistic fuzzy metric space which satisfies the following conditions:*
$$M(x,y,0) = \lim_{t \to 0} M(x,y,t) = 0$$
and
$N(x,y,0) = \lim_{t \to 0} N(x,y,t) = 1$ *for $x \neq y$.*

Further, let $\varphi : (0, \infty) \to (0, \infty)$ be a continuous, non-decreasing function which satisfies $\varphi(t) < t$ for all $t > 0$. Then the following statements hold:

(a) if $M(x, y, \varphi(t)) \geq M(x,y,t)$ for all $t > 0$, then $x = y$,
(b) if $N(x, y, \varphi(t)) \leq N(x,y,t)$ for all $t > 0$, then $x = y$.

Alaca et al. [3] proved the following result:

THEOREM A. (Instuitionistic fuzzy Banach contraction theorem). *Let $(X, M, N, *, \diamond)$ be a complete intuitionistic fuzzy metric space. Let $T : X \to X$ be a mapping satisfying*

$$M(Tx, Ty, kt) \geq M(x,y,t) \text{ and } N(Tx, Ty, kt) \leq N(x,y,t)$$

for all x, y in X where $0 < k < 1$. Then T has a unique fixed point.

To prove our results we need the following Lemma:

Lemma 2.2. [1, 2]. *Let $\{y_n\}$ be a sequence in an intuitionistic fuzzy metric space $(X, M, N, *, \diamond)$ such that $b*b \geq b$ and $(1-b) \diamond (1-b) \leq (1-b)$ for all $b \in (0, 1)$. If there exists a constant $k \in (0, 1)$ such that*

(2.1)
$$M(y_n, y_{n+1}, kt) \geq M(y_{n-1}, y_n, t) \text{ and } N(y_n, y_{n+1}, kt) \leq N(y_{n-1}, y_n, t)$$

for all $n \in N$, where $t > 0$, then $\{y_n\}$ is a Cauchy sequence in X.

3. COINCIDENCES AND FIXED POINT THEOREMS

Theorem 3.1. *Let Y be an arbitrary set and $(X, M, N, *, \diamond)$ be an intuitionistic fuzzy metric space, such that $b*b \geq b$ and $(1-b) \diamond (1-b) \leq (1-b)$ for all $b \in (0, 1)$. Let $\{A_i\}_{i \in N} : Y \to X$, $B : X \to Y$ and mappings $P, Q, S, T : Y \to X$ be such that*
  (a) $A_i(Y) \subset PBQ(Y) \cap SBT(Y)$, $i \in N$.
  (b) *there exists a constant $k \in (0, 1)$ such that*

$$\int_0^{M(A_i x, A_j y, kt)} \varphi(t) dt \geq \int_0^{m(x, y, t)} \varphi(t) dt$$

and

$$\int_0^{N(A_i x, A_j y, kt)} \varphi(t) dt \geq \int_0^{n(x, y, t)} \varphi(t) dt,$$

where $\varphi : R^+ \to R^+$ is a Lebesgue integrable mapping which is summable on each compact subset of R^+, non-negative and such that

$$\int_0^\epsilon \varphi(t) dt > 0 \text{ for each } \epsilon > 0,$$

where
$$m(x, y, t) \geq M(A_i x, PBQx, t) * M(A_j y, SBTy, t)$$

(3.1) $\quad * M(A_i x, SBTy, \alpha t) * M(A_j y, PBQx, (2-\alpha)t)$

$$n(x, y, t) \leq N(A_i x, PBQx, t) \diamond N(A_j y, SBTy, t)$$

(3.2) $\quad \diamond N(A_i x, SBTy, \alpha t) \diamond N(A_j y, PBQx, (2-\alpha)t),$

for all $t > 0$ and $\alpha \in (0, 2)$ and for each $x, y \in Y, i, j \in N$ with $i \neq j$,

(3.3) $\quad PBQ(Y) \cap SBT(Y)$ is a complete subspace of X.

Then for each $i \in N$,
(i) A_i and PBQ have a coincidence point,
(ii) A_i and SBT have a coincidence point.

PROOF. Choose a point $x_0 \in Y$. Since $A_i(Y) \subset SBT(Y)$, we can choose a point $x_1 \in Y$ such that $A_1 x_0 = SBT x_1$.

Similarly, we can choose a point $x_2 \in Y$ such that $A_2 x_1 = PBQ x_2$. Inductively, we can find a sequence $\{x_n\}$ in Y such that

$$A_{2n} x_{2n-1} = PBQ x_{2n} = y_{2n}$$

and

$$A_{2n+1} x_{2n} = SBT x_{2n+1} = y_{2n+1}.$$

Then by (b), with $\alpha = 1 + q$ and $q \in (0, 1)$, we have

$$\int_0^{M(y_{2n+1}, y_{2n}, kt)} \varphi(t) dt = \int_0^{M(A_{2n+1} x_{2n}, A_{2n} x_{2n-1}, kt)} \varphi(t) dt$$
$$\geq \int_0^{m(x_{2n}, x_{2n-1}, t)} \varphi(t) dt.$$

Now, using (3.1), we have

$m(x_{2n}, x_{2n-1}, t) \geq M(A_{2n+1} x_{2n}, PBQ x_{2n}, t)$
$\quad * M(A_{2n} x_{2n-1}, SBT x_{2n-1}, t)$
$\quad * M(A_{2n+1} x_{2n}, SBT x_{2n-1}, (1+q)t)$
$\quad * M(A_{2n} x_{2n-1}, PBQ x_{2n}, (1-q)t)$
$\quad = M(y_{2n+1}, y_{2n}, t) * M(y_{2n}, y_{2n-1}, t)$
$\quad * M(y_{2n+1}, y_{2n-1}, (1+q)t) * M(y_{2n}, y_{2n}, (1-q)t)$
$\quad \geq M(y_{2n+1}, y_{2n}, t) * M(y_{2n}, y_{2n-1}, t)$
$\quad * M(y_{2n+1}, y_{2n}, t) * M(y_{2n}, y_{2n-1}, qt)$
$\quad * M(y_{2n}, y_{2n-1}, t) * 1$
$\quad \geq M(y_{2n+1}, y_{2n}, t) * M(y_{2n}, y_{2n-1}, t)$

and then

$$\int_0^{N(y_{2n+1},y_{2n},kt)} \varphi(t)dt = \int_0^{N(A_{2n+1}x_{2n},A_{2n}x_{2n-1},kt)} \varphi(t)dt$$
$$= \int_0^{n(x_{2n},x_{2n-1},t)} \varphi(t)dt.$$

Now, using (3.2), we have

$$\begin{aligned} n(x_{2n},x_{2n-1},t) &\leq N(y_{2n+1},y_{2n},t) \diamond N(y_{2n},y_{2n-1},t) \\ &\diamond N(y_{2n+1},y_{2n},t) \diamond N(y_{2n},y_{2n-1},qt) \\ &\diamond N(y_{2n},y_{2n-1},t) \diamond 0 \\ &= N(y_{2n+1},y_{2n},t) \diamond N(y_{2n},y_{2n-1},t). \end{aligned}$$

Since the t-norm $*$ and the t-norm $\diamond$ are continuous, M(x, y, •) is left continuous and N(x, y, •) is right continuous, we get on letting $q \to 1$,

$$m(x_{2n},x_{2n-1},t) \geq M(y_{2n},y_{2n-1},t) * M(y_{2n+1},y_{2n},t) N(y_{2n+1},y_{2n},t)$$
$$\diamond N(y_{2n},y_{2n-1},t)$$

and

$$n(x_{2n},x_{2n-1},t) \leq N(y_{2n},y_{2n-1},t) \diamond N(y_{2n+1},y_{2n},t).$$

Therefore

$$\int_0^{M(y_{2n+1},y_{2n},kt)} \varphi(t)dt \geq \int_0^{M(y_{2n},y_{2n-1},t)*M(y_{2n+1},y_{2n},t)} \varphi(t)dt$$

and

$$\int_0^{N(y_{2n+1},y_{2n},kt)} \varphi(t)dt \leq \int_0^{N(y_{2n},y_{2n-1},t)\diamond N(y_{2n+1},y_{2n},t)} \varphi(t)dt.$$

By repeated applications of the above argument, we have

$$\int_0^{M(y_{m+2},y_{m+1},kt)} \varphi(t)dt \geq \int_0^{M(y_{m+1},y_m,t)*M(y_{m+2},y_{m+1},tk^{-p})} \varphi(t)dt$$

and

$$\int_0^{N(y_{m+2},y_{m+1},kt)} \varphi(t)dt \leq \int_0^{N(y_{m+1},y_m,t)\diamond N(y_{m+2},y_{m+1},tk^{-p})} \varphi(t)dt$$

for all $m, p \in N$.

Since $M(y_{m+2}, y_{m+1}, tk^{-p}) \to 1$ and $N(y_{m+2}, y_{m+1}, tk^{-p}) \to 0$ as $p \to \infty$, we obtain

$$\int_0^{M(y_{m+2},y_{m+1},kt)} \varphi(t)dt \geq \int_0^{M(y_{m+1},y_m,t)} \varphi(t)dt$$

and

$$\int_0^{N(y_{m+2},y_{m+1},kt)} \varphi(t)dt \leq \int_0^{N(y_{m+1},y_m,t)} \varphi(t)dt$$

for all $m \in N$.

By Lemma 2.7, $\{y_n\}$ is a Cauchy sequence in $PBQ(Y) \cap SBT(Y)$ and so has a limit, say u. Then there exist v and w such that

$$v \in (PBQ)^{-1}u \text{ and } w \in (SBT)^{-1}u.$$

Therefore, $PBQv = u = SBTw$.

We will now prove that $A_i v = u$. By (b) and (3.1), with $\alpha = 1+k$, we have

$$\int_0^{M(A_i v, y_{2n}, t)} \varphi(t)dt = \int_0^{M(A_i v, A_{2n} x_{2n-1}, t)} \varphi(t)dt$$
$$= \int_0^{m(v, x_{2n-1}, t)} \varphi(t)dt.$$

Now by (3.1), we have

$$\begin{aligned}
m(v, x_{2n-1}, t) &\geq M(A_i v, PBQv, t/k) \\
&\quad * M(A_{2n} x_{2n-1}, SBT x_{2n-1}, t/k) \\
&\quad * M(A_i v, SBT x_{2n-1}, (1+k)t/k) \\
&\quad * M(A_{2n} x_{2n-1}, PBQv, (1-k)t/k) \\
&= M(A_i v, PBQv, t/k) * M(y_{2n}, y_{2n-1}, t/k) \\
&\quad * M(A_i v, y_{2n-1}, (1+k)t/k) \\
&\quad * M(y_{2n}, PBQv, (1-k)t/k).
\end{aligned}$$

Similarly from (b) and (3.2), we obtain

$$\begin{aligned}
n(v, x_{2n-1}, t) &\leq N(A_i v, PBQv, t/k) \diamond N(y_{2n}, y_{2n-1}, t/k) \\
&\quad \diamond N(A_i v, y_{2n-1}, (1+k)t/k) \\
&\quad \diamond N(y_{2n}, PBQv, (1-k)t/k).
\end{aligned}$$

Since $\{y_n\}$ is a Cauchy sequence, it converges to $u = PBQv$ and so on letting $n \to \infty$, we have

$$m(v, x_{2n-1}, t) \geq M(A_iv, PBQv, t/k) * 1$$
$$* M(A_iv, PBQv, (1+k)t/k) * 1$$
$$\geq M(A_iv, PBQv, (1+k)t/k) \text{ and}$$
$$n(v, x_{2n-1}, t) \leq N(A_iv, PBQv, (1+k)t/k).$$

Therefore

$$\int_0^{M(A_iv, PBQv, t)} \varphi(t)dt \geq \int_0^{M(A_iv, PBQv, (1+k)t/k)} \varphi(t)dt$$

and

$$\int_0^{N(A_iv, PBQv, t)} \varphi(t)dt \leq \int_0^{N(A_iv, PBQv, (1+k)t/k)} \varphi(t)dt.$$

Since $M(x, y, \bullet)$ is non-decreasing and $N(x, y, \bullet)$ is non-increasing, we must have

$$A_iv = PBQv = u$$

for each $i \in N$.

Similarly

$$A_iw = SBTw = u$$

for each $i \in N$.

Thus, v is a coincidence point of A_i and PBQ and w is a coincidence point of A_i and SBT. This completes the proof of the theorem.

In the following, $C(S, T)$ stands for the set of coincidence points of the mappings S and T i.e.

$$C(S, T) = \{z \in X : Sz = Tz\}.$$

Theorem 3.2. Let $(X, M, N, *, \diamond)$ be an intuitionistic fuzzy metric space with $b * b \geq b$ and $(1-b) \diamond (1-b) \leq (1-b)$ for all $b \in (0,1)$. Let $\{A_i\}$ be a family of self mappings on X. Suppose that P, Q, S, T be self mappings on X such that $A_i(X) \subset PQ(X) \cap ST(X)$ for all $i \in N$ and conditions (2.2) and (2.3) hold. Suppose further that A_i commutes with each of P, Q, S and T; PQ commutes with S and T; P commutes with Q. Then P, Q, S, T and the family $\{A_i\}$ have a unique common fixed point.

PROOF. In Theorem 3.1, we have proved that $v \in C(A_i, PQ)$ and $w \in C(A_i, ST)$. We then notice that

(3.4) $$PQv = A_i v = u = STw = A_j w,$$

(3.5) $$A_i v = A_i PQv = PA_i Qv = PQA_i v = PQu$$

and

(3.6) $$A_j u = A_j STw = Sa_j Tw = STA_j w = STu.$$

Now from (b) and (3.1), taking $\alpha = 1$ and using (2.5) - (2.7), we have

$$\int_0^{M(u, A_i u, kt)} \varphi(t) dt = \int_0^{M(A_i v, A_j u, kt)} \varphi(t) dt$$

$$= \int_0^{m(v, u, t)} \varphi(t) dt$$

$$\geq \int_0^{M(A_i v, PQv, t) * M(A_j u, STu, t) * M(A_i v, STu, t) * (M(A_j u, PQv, t)} \varphi(t) dt$$

$$= \int_0^{1 * 1 * M(u, A_j u, t) * M(A_j u, u, t)} \varphi(t) dt$$

$$\geq \int_0^{M(u, A_i u, t)} \varphi(t) dt$$

and from (3.2), we have

$$\int_0^{N(A, A_i u, kt)} \varphi(t) dt \leq \int_0^{M(u, A_i u, t)} \varphi(t) dt,$$

which implies that $u = A_j u$.

From (3.5) and (3.6), u is a common fixed point of PQ and ST.

We now show that u is a common fixed point of P, Q, S and T. From (3.1) and using (3.4) - (3.6), we have

$$\int_0^{M(Su, u, kt)} \varphi(t) dt = \int_0^{M(SA_i v, A_j u, kt)} \varphi(t) dt$$

$$= \int_0^{M(A_i Sv, A_j u, kt)} \varphi(t) dt$$

$$\geq \int_0^{M(A_iSv,PQSv,t)*M(A_ju,STu,t)*M(A_iSv,STu,\alpha t)*(M(A_ju,PQSv,(2-\alpha)t)} \varphi(t)dt$$

$$= \int_0^{M(SA_iv,SPQv,t)*M(A_ju,A_ju,t)*M(SA_iv,STu,\alpha t)*(M(A_ju,SPQv,(2-\alpha)t)} \varphi(t)dt$$

$$= \int_0^{M(Su,Su,t)*M(u,u,t)*M(Su,u,\alpha t)*(M(u,Su,(2-\alpha)t)} \varphi(t)dt$$

$$\geq \int_0^{1*1*M(Su,u,\alpha t)} \varphi(t)dt$$

$$\geq \int_0^{M(Su,u,\alpha t)} \varphi(t)dt.$$

Similarly,

$$\int_0^{N(Su,u,kt)} \varphi(t)dt \leq \int_0^{N(Su,u,\alpha t)} \varphi(t)dt.$$

and letting $\alpha \to 1$, we have $Su = u$.

Similarly, we can prove that $Tu = u = Pu = Qu$. Hence u is a common fixed point of P, Q, S and T.

Finally, we prove that u is the unique common fixed point. Suppose that u and v are two distinct fixed points. Taking $\alpha = 1$ from (b) and (3.1), we have

$$\int_0^{M(u,v,kt)} \varphi(t)dt = \int_0^{M(A_iu,A_jv,kt)} \varphi(t)dt$$

$$\geq \int_0^{M(A_iu,PQu,t)*M(A_jv,STv,t)*M(A_iu,STv,t)*(M(A_jv,PQu,t)} \varphi(t)dt$$

$$= \int_0^{1*1*M(u,v,t)*M(v,u,t)} \varphi(t)dt$$

$$\geq \int_0^{M(u,v,t)} \varphi(t)dt$$

and

$$\int_0^{N(u,v,kt)} \varphi(t)dt \leq \int_0^{N(u,v,t)} \varphi(t)dt,$$

giving a contradiction which implies that $u = v$. Hence u is the unique common fixed point. This completes the proof of the theorem.

Corollary 3.1. *Let Y be an arbitrary set, let X an intuitionistic FM-space and let $A, B : Y \to X$. If there exists a constant $k \in (0, 1)$ and mappings $P, Q, S, T : Y \to X$ such that*
(a) $A(Y) \cup B(Y) \subset PQ(Y) \cap ST(Y)$, and for each $x, y \in Y$,
(b) there exists a constant $k \in (0, 1)$, such that

$$\int_0^{M(Ax,By,kt)} \varphi(t)dt \geq \int_0^{m(x,y,t)} \varphi(t)dt,$$

$$\int_0^{N(Ax,By,kt)} \varphi(t)dt \geq \int_0^{n(x,y,t)} \varphi(t)dt,$$

for all $x, y \in X$, where $\varphi : R^+ \to R^+$ is a Lebesgue integrable mappings which is summable on each compact subset of R^+, non-negative and such that

$$\int_0^{\epsilon} \varphi(t)dt > 0 \text{ for each } \epsilon > 0,$$

(3.7)
$$m(Ax, By, kt) \geq M(Ax, PQx, t) * M(By, STy, t) * M(Ax, STy, \alpha t)$$
$$*M(By, PQx, (2-\alpha)t)$$

(3.8)
$$N(Ax, By, kt) \leq N(Ax, PQx, t) \diamond N(By, STy, t) \diamond N(Ax, STy, \alpha t)$$
$$\diamond N(By, PQx, (2-\alpha)t)$$

for all $t > 0$ and $\alpha \in (0, 2)$;

(3.9) $PQ(Y) \cap ST(Y)$ *is a complete subspace of X;*

(i) then A and PQ have a coincidence point,
(ii) B and ST have a coincidence point.
Further, if $X = Y$, and

$$PQAu = PAQu = APQu; \ u \in C(A, PQ)$$

and

$$BSTv = SBTv = STBv; \ v \in C(B, ST),$$

then A, B, PQ and ST have a unique common fixed point.

This corollary is indeed Theorem 3.1 and Theorem 3.2 together with $A = A_{2i+1}$ and $B = A_{2i}$, $i \in N$.

Corollary 3.2. *Let $A, B, ST : Y \to X$. If $A(Y) \cap B(Y) \subset ST(Y)$ and (3.7)-(3.9) hold with $PQ = ST$, then A, B and ST have a coincidence point, i.e., there exists a point z in Y such that $Az = Bz = STz$. Further, if $X = Y$ and ST commutes with each of A and B (only) at z, then A, B and ST have a unique common fixed point.*

Corollary 3.3. *Let X be a complete intuitionistic FM-space and $A, B : X \to X$. If there exists a constant $k \in (0,1)$ such that*

$$\int_0^{M(Ax,By,kt)} \varphi(t)dt \geq \int_0^{m(x,y,t)} \varphi(t)dt,$$

$$\int_0^{N(Ax,By,kt)} \varphi(t)dt \geq \int_0^{n(x,y,t)} \varphi(t)dt,$$

for all $x, y \in X$, where $\varphi : R^+ \to R^+$ is a Lebesgue integrable mapping which is summable on each compact subset of R^+, non-negative and such that

$$\int_0^\epsilon \varphi(t)dt > 0 \ for \ each \ \epsilon > 0,$$

$$m(Ax, By, kt) \geq M(x, Ax, t) * M(y, By, t)$$
$$* M(y, Ax, \alpha t) * M(x, By, (2-\alpha)t)$$
$$m(Ax, By, kt) \leq N(x, Ax, t) \diamond N(y, By, t)$$
$$\diamond N(y, Ax, \alpha t) \diamond N(x, By, (2-\alpha)t)$$

for all $t > 0$ and $\alpha \in (0,2)$, then A and B have a unique common fixed point.

Remark 3.1. *A number of coincidence and common fixed point theorems may be obtained as the special cases of Theorem 3.1 and Theorem 3.2 for two to four mappings in intuitionistic FM-spaces.*

REFERENCES

[1] C. ALACA, Fixed point theorems for weak compatible mappings in intuitionistic fuzzy metric spaces, *Int. J. Pure Appl. Math.*, **32**(4) (2006) 537-548.

[2] C. ALACA, I. ALTUN AND D. TURKOGLU, On compatible mappings of type (I) and (II) in intuitionistic fuzzy metric spaces, *Commun. Korean Math. Soc.*, **23(2)** (2009) 427-446.

[3] C. ALACA, D. TURKOGLU AND C. YILDIZ, Fixed points in intuitionistic fuzzy metric spaces, *Chaos, Solitons & Fractals*, **29** (2006) 1073-1078.

[4] K. ATANASSOV, Intuitionistic fuzzy sets, *Fuzzy Sets & Systems*, **20**(1986) 87-96.

[5] K. ATANASSOV, New operations defined over the intuitionistic fuzzy sets, *Fuzzy Sets & Systems*, **61** (1994) 137-142.

[6] G. DESCHRIJVER, C. CORNELIS AND E.E. KERRE, On the representation of intuitionistic fuzzy t-norms and t-conorms, *IEEE Trans. Fuzzy Syst.*, **12**(2004) 45-61.

[7] G. DESCHRIJVER AND E.E. KERRE, On the relationship between some extensions of fuzzy set theory, *Fuzzy Sets & Syst.*, **33**(2003) 227-235.

[8] R.C. DIMRI and N.S. GARIYA, Coincidences and common fixed points in intuitionistic fuzzy metric spaces, *Indian J. Math.*, **52 (3)** (2010), 479–490.

[9] M. GRABIEC, Fixed points in fuzzy metric spaces, *Fuzzy Sets & Systems*, **27**(1988) 385-389.

[10] S.N. JESIC AND N.A. BABACEV, Common fixed point theorems in intuitionistic fuzzy metric spaces and ℓ-fuzzy metric spaces with nonlinear contractive condition, *Chaos, Solitons & Fractals*, **37 (3)** (2008) 675-687.

[11] S.N. MISHRA, N. SHARMA AND S.L. SINGH, Common fixed point of maps on fuzzy metric spaces, *Internat J. Math. Math. Sci.*, **17** (1994) 253-258.

[12] S.N. MISHRA, S.L. SINGH AND V. CHADHA, Coincidences and fixed points in fuzzy metric spaces, *The J. Fuzzy Maths.*, **6 (2)** (1998) 491-500.

[13] J.H. PARK, Intuitionistic fuzzy metric spaces, *Chaos, Solitons & Fractals*, **22** (2004) 1039-1046.

[14] B. SCHWEIZER AND A. SKLAR, Statistical metric spaces, *Pacific J. Math.*, **10** (1960) 313-334.

[1] DEPARTMENT OF MATHEMATICS,
GOVERNMENT DEGREE COLLEGE BILLAWAR,
JAMMU AND KASHMIR, INDIA-184204
E-mail address: singhamit841@gmail.com

[2] DEPARTMENT OF MATHEMATICS,
UNIVERSITY OF LEICESTER,
LEICESTER, LE1 7RH, ENGLAND.
E-mail address: fbr@le.ac.uk

Differential Geometry, Functional Analysis and Applications
Editors: Mohammad Hasan Shahid, Sharfuddin Ahmad et al.
Copyright © 2015, Narosa Publishing House, New Delhi

CHARACTERIZATION OF SOBOLEV SPACES USING M-BAND FRAMELET PACKETS

F. A. SHAH*, HUMAIRA SIDDIQUI** AND K. AHMAD**

ABSTRACT. In [Shah and Debnath, Explicit construction of M-band framelet packets, *Analysis*, 32 (2012), 281-294], authors have given a general construction scheme for a class of stationary M-band tight framelet packets in $L^2(\mathbb{R})$ via extension principles. In this paper, we use these M-band framelet packets to characterize the Sobolev norm of any function $f \in \mathbb{H}^s(\mathbb{R})$, $-\alpha < s < \alpha$ by means of the framelet packet coefficient sequence $\left\{\langle f, \omega_{n,j,k}\rangle\right\}_{I_{j,n}\in\Gamma_J, k\in\mathbb{Z}} \cup \left\{\langle f, \psi_{\ell,j,k}\rangle\right\}_{\ell=1,\ldots,L, j\geq J, k\in\mathbb{Z}}$.

[1]

1. INTRODUCTION

The traditional wavelet frames provide poor frequency localization in many applications as they are not suitable for signals whose domain frequency channels are focused only on the middle frequency region. Therefore, in order to make more kinds of signals suited for analyzing by wavelet frames, it is necessary to extend the concept of wavelet frames to a library of wavelet frames, called *framelet packets* or *wavelet frame packets*. The original idea of framelet packets was introduced by Coifman *et al.* in [3] to provide more efficient decomposition of signals containing both transient and stationary components. The concept of wavelet packet was further generalized to many different setups, for example, Chui and

---

[1]*2000 Mathematics Subject Classification:* 42C40; 42C15; 65T60

Keywords: M-band wavelets; Tight wavelet frame; Framelet packet; Extension principle; Fourier transform; Sobolev spaces

Li [2] generalized the concept of orthogonal wavelet packets to the case of non-orthogonal wavelet packets so that they can be applied to the spline wavelets and so on. In his recent paper, Shah [10] has constructed p-wavelet packets on the positive half-line $\mathbb{R}^+$ using the classical splitting trick of wavelets where as Shah and Debnath in [11] have constructed the corresponding p-wavelet frame packets on $\mathbb{R}^+$ using the Walsh-Fourier transform. The introduction of M-band wavelet packets on $\mathbb{R}$ attributes to Jiankang et al. [6].

On the otherhand, the standard orthogonal wavelets are not also suitable for the analysis of high-frequency signals with relatively narrow bandwidth. To overcome this shortcoming, M-band orthonormal wavelets were created as a direct generalization of the 2-band wavelets [13]. The motivation for a larger $M(M > 2)$ comes from the fact that, unlike the standard wavelet decomposition which results in a logarithmic frequency resolution, the M-band decomposition generates a mixture of logarithmic and linear frequency resolution and hence generates a more flexible tiling of the time-frequency plane than that resulting from 2-band wavelet. The other significant difference between 2-band wavelets and M-band wavelets in construction lies in the aspect that the wavelet vectors are not uniquely determined by the scaling vector and the orthonormal bases do not consist of dilated and shifted functions through a single wavelet, but consist of ones by using $M-1$ wavelets (see [1,7]). It is this point that brings more freedoms for optimal wavelet bases.

A tight wavelet frame is a generalization of an orthonormal wavelet basis by introducing redundancy into a wavelet system. Tight wavelet frames have some desirable features such as near translation invariant wavelet frame transforms and it may also be easier to recognize patterns in a redundant transform. A catalyst for this development is the unitary extension principle (UEP) introduced by Ron and Shen [9], which provides a general construction of tight wavelet frames for $L^2(\mathbb{R}^n)$ in the shift-invariant setting, and included the pyramidal decomposition and reconstruction filter bank algorithms. The resulting tight wavelet frames are based on a multiresolution analysis, and the generators are often called *mother framelets*. The theory of tight wavelet frames has been extensively studied and well developed over the recent years. In the M-band setting, Han and Cheng [5] have provided the general construction of M-band tight wavelet frames on $\mathbb{R}$ by following the procedure of Daubechies *et al.*[4] and Petukhov [8] via extension principles.

Recently, Shah and Debnath [12] have introduced a general construction scheme for a class of *stationary M-band tight framelet packets* in $L^2(\mathbb{R})$ via extension principles. They proved a lemma on the so-called splitting trick and splited the wavelet spaces $W_{j,\ell}, \ell = 0, 1, ..., L$ by means of the framelet symbols $m_\ell, \ell = 0, 1, ..., L$ and then by recursive decomposition, constructed various M-band tight framelet packets in $L^2(\mathbb{R})$. In this paper, we use the weighted l^2-norm of the stationary M-band framelet packet coefficient sequence $\{\langle f, \omega_{n,j,k}\rangle\}_{I_{j,n} \in \Gamma_J, k \in \mathbb{Z}} \cup \{\langle f, \psi_{\ell,j,k}\rangle\}_{\ell=1,...,L, j \geq J, k \in \mathbb{Z}}$ of a given function $f \in \mathbb{H}^s(\mathbb{R})$ to characterize its Sobolev norm in $\mathbb{H}^s(\mathbb{R}), -\alpha < s < \alpha$.

The rest of this paper is organized as follows. In Section 2, we review some basic facts about stationary M-band framelet packets in $L^2(\mathbb{R})$ using extension principles. In Sections 3, we prove our main result regarding the characterization of Sobolev spaces $\mathbb{H}^s(\mathbb{R}), -\alpha < s < \alpha$ using stationary M-band framelet packets.

2. PRELIMINARIES AND M-BAND FRAMELET PACKETS

We begin this section by reviewing some major concepts concerning M-band framelet packets. In the rest of this paper, we use $\mathbb{N}, \mathbb{Z}$ and $\mathbb{R}$ to denote the sets of all natural numbers, integers and real numbers, respectively.

The Fourier transform of a function $f \in L^1(\mathbb{R})$ is defined as usual by:

$$\hat{f}(\xi) = \int_\mathbb{R} f(x) e^{-i\xi x} dx, \quad \xi \in \mathbb{R}$$

and its inverse is

$$f(x) = \frac{1}{2\pi} \int_{\mathbb{R}} \hat{f}(\xi) e^{i\xi x} d\xi, \quad x \in \mathbb{R}.$$

For a real number s, we denote by $\mathbb{H}^s(\mathbb{R})$ the Sobolev space consisting of all tempered distributions f such that

$$\|f\|_{\mathbb{H}^s(\mathbb{R})}^2 = \frac{1}{2\pi} \int_{\mathbb{R}} \left|\hat{f}(\xi)\right|^2 \left(1 + |\xi|^2\right)^s d\xi < \infty.$$

Note that $\mathbb{H}^0(\mathbb{R}) = L^2(\mathbb{R})$ and $\|f\|_{\mathbb{H}^0(\mathbb{R})} = \|f\|_{L^2(\mathbb{R})}$ by the Plancherels theorem.

For $f, g \in L^2(\mathbb{R})$, we define the *bracket product* function $[.,.]$ as

$$[f,g] = \sum_{k \in \mathbb{Z}} f(. + 2\pi k) \overline{g(. + 2\pi k)}.$$

Clearly $[.,.] \in L^1(\mathbb{T})$ whenever $f, g \in L^2(\mathbb{R})$, where $\mathbb{T}$ is any tour of $\mathbb{R}$. Moreover, for $f, g \in L^2(\mathbb{R})$, $[.,.]_s$ is defined as

$$(2.1) \quad [f,g]_s = \sum_{k \in \mathbb{Z}} f(. + 2\pi k) \overline{g(. + 2\pi k)} \left(1 + |. + 2\pi k|^2\right)^s d\xi.$$

For given $\Psi := \{\psi_1, \ldots, \psi_L\} \subset L^2(\mathbb{R})$, define the M-band wavelet system

$$X(\Psi) := \left\{\psi_{\ell,j,k} : 1 \leq \ell \leq L; j, k \in \mathbb{Z}\right\}$$

where $\psi_{\ell,j,k} = M^{j/2} \psi_\ell(M^j . - k)$. The wavelet system $X(\Psi)$ is called a M-band wavelet frame, or simply a M-band framelet system, if there exist positive numbers $0 < A \leq B < \infty$ such that for all $f \in L^2(\mathbb{R})$

$$(2.2) \quad A\|f\|^2 \leq \sum_{\ell=1}^{L} \sum_{j \in \mathbb{Z}} \sum_{k \in \mathbb{Z}} |\langle f, \psi_{\ell,j,k} \rangle|^2 \leq B\|f\|^2.$$

The largest A and the smallest B for which (2.2) holds are called wavelet frame bounds. A wavelet frame is a *tight wavelet frame* if A and B are chosen such that $A = B = 1$ and then generators $\psi_1, \psi_2, ..., \psi_L$ are often referred as *M-band framelets*.

The construction of framelet systems often starts with the construction of MRA, which is built on refinable functions. A function $\varphi \in L^2(\mathbb{R})$ is called *M-refinable* if it satisfies a refinement equation:

$$(2.3) \qquad \varphi(x) = \sum_{k \in \mathbb{Z}} h_0[k]\, \varphi(Mx - k),$$

for some $h_0 \in l^2(\mathbb{Z})$. The Fourier transform of (2.3) yields

$$(2.4) \qquad \hat{\varphi}(\xi) = m_0\left(\frac{\xi}{M}\right) \hat{\varphi}\left(\frac{\xi}{M}\right),$$

where

$$m_0(\xi) = \frac{1}{M} \sum_{k \in \mathbb{Z}} h_0[k] e^{ik\xi},$$

is a 2π-periodic measurable function in $L^\infty[-\pi, \pi]$ and is often called the *refinement symbol* of φ. Given an M-refinable function $\varphi \in L^2(\mathbb{R})$ with $\hat{\varphi}(0) \neq 0$, the sequence of subspaces $\{V_j\}_{j \in \mathbb{Z}}$ defined by

$$V_j = \overline{\mathrm{span}}\left\{\varphi(M^j x - k) : k \in \mathbb{Z}\right\}, \quad j \in \mathbb{Z}$$

will form an MRA for $L^2(\mathbb{R})$. Recall that $\{V_j\}_{j \in \mathbb{Z}}$ is called an MRA if it satisfies (i) $V_j \subset V_{j+1}$ for every $j \in \mathbb{Z}$; (ii) $\bigcup_{j \in \mathbb{Z}} V_j$ is dense in $L^2(\mathbb{R})$ and (iii) $\bigcap_{j \in \mathbb{Z}} V_j = \{0\}$. In this paper, we only consider the refinable function $\varphi \in L^2(\mathbb{R})$ satisfying the following properties:

$$(2.5) \qquad \lim_{\xi \to 0} \hat{\varphi}(\xi) = 1, \quad \xi \in \mathbb{R};$$

and

$$(2.6) \qquad \sum_{k \in \mathbb{Z}} |\hat{\varphi}(\xi + 2k\pi)|^2 \in L^\infty[-\pi, \pi].$$

Given an MRA generated by the M-refinable function φ, one can construct (see [4]) a set of MRA-based framelets $\Psi := \{\psi_1, ..., \psi_L\} \subset V_1$ which is defined by

$$\hat{\psi}_\ell(\xi) = m_\ell\left(\frac{\xi}{M}\right) \hat{\varphi}\left(\frac{\xi}{M}\right), \tag{2.7}$$

where

$$m_\ell(\xi) = \frac{1}{M} \sum_{k \in \mathbb{Z}} h_\ell[k] e^{ik\xi}, \quad \ell = 1, \ldots, L$$

are the 2π-periodic measurable functions in $L^\infty[-\pi, \pi]$ and are called the *framelet symbols* or *wavelet masks*. The so-called *unitary extension principle* (UEP) provides a sufficient condition on Ψ such that the resulting M-band system $X(\Psi)$ forms a tight frame of $L^2(\mathbb{R})$. In this connection, an explicit construction scheme is provided in [5] for the construction of M-band tight framelets on $\mathbb{R}$.

Theorem 2.1[5]. Suppose that the refinable function φ and the framelet symbols $m_0, m_1, \ldots, m_L$ satisfy (2.4)–(2.6). Define $\psi_1, \ldots, \psi_L$ by (2.7). Let $\mathcal{M}(\xi) = \left\{m_\ell\left(\xi + \frac{2\pi p}{M}\right)\right\}_{\ell, p=0}^{M-1}$ such that $\mathcal{M}(\xi)\mathcal{M}^*(\xi) = I_M$, for a.e $\xi \in \sigma(V_0) := \left\{\xi \in [-\pi, \pi] : \sum_{k \in \mathbb{Z}} |\hat{\varphi}(\xi + 2k\pi)|^2 \neq 0\right\}$, then M-band wavelet system $X(\Psi)$ forms a tight wavelet frame for $L^2(\mathbb{R})$ with frame bound 1.

Let $X(\Psi)$ be the M-band tight wavelet frame for $L^2(\mathbb{R})$ constructed via UEP in an MRA $\{V_j\}_{j \in \mathbb{R}}$ generated by the M-refinable function φ with combined UEP mask $\mathbf{m} = [m_0, m_1, \ldots, m_L]$. Then, for each $j \in \mathbb{Z}$, we define

$$V_j = \overline{\text{span}}\{\varphi_{j,k} : k \in \mathbb{Z}\},$$

and

$$W_{j,\ell} = \overline{\text{span}}\{\psi_{\ell,j,k} : k \in \mathbb{Z}\}, \quad \ell = 0, 1, \ldots, L.$$

Therefore, in view of tight frame decomposition, we have

$$V_j = V_{j-1} + \sum_{\ell=1}^{L} W_{j-1,\ell}. \tag{2.8}$$

It is immediate from the above decomposition that these $L + 1$ spaces are in general not orthogonal. Therefore, by the repeated applications of (2.8), we can further split the V_j spaces as:

$$V_j = V_{j-1} + \sum_{\ell=1}^{L} W_{j-1,\ell} = V_{j-2} + \sum_{r=j-2}^{j-1} \sum_{\ell=1}^{L} W_{r,\ell} = \ldots\ldots$$

$$= V_{j_0} + \sum_{r=j_0}^{j-1} \sum_{\ell=1}^{L} W_{r,\ell} = \sum_{r=-\infty}^{j-1} \sum_{\ell=1}^{L} W_{r,\ell}.$$

Recently, Shah and Debnath [12] have constructed various *stationary tight M-band framelet packets* on $\mathbb{R}$ by the recursive decomposition of wavelet spaces $W_{j,\ell}, \ell = 0, 1, \ldots, L, j \in \mathbb{Z}$.

For $n = 0, 1, 2, \ldots$, the *basic M-band framelet packets* associated with the M-refinable function φ are defined as

$$\hat{\omega}_n(\xi) = \hat{\omega}_{(L+1)r+\ell}(\xi) = m_\ell\left(\frac{\xi}{M}\right) \hat{\omega}_r\left(\frac{\xi}{M}\right),$$

(2.9) $\qquad \ell = 0, 1, \ldots, L, r = 0, 1, 2, \ldots$

Note that for $r = 0$ and $\ell = 1, \ldots, L$, we have

$$\hat{\omega}_\ell(\xi) = m_\ell\left(\frac{\xi}{M}\right) \hat{\omega}_0\left(\frac{\xi}{M}\right) = m_\ell\left(\frac{\xi}{M}\right) \hat{\varphi}\left(\frac{\xi}{M}\right),$$

which shows that $\omega_\ell(x) = \psi_\ell(x), \ell = 1, \ldots, L$.

Define a family of subspaces of $L^2(\mathbb{R})$ by

(2.10) $\qquad U_n := \overline{\text{span}}\{\omega_{n,0,k} : k \in \mathbb{Z}\}, \quad n = 0, 1, 2, \ldots$

Clearly $U_0 = V_0$ and $U_\ell = W_{0,\ell}$ for $\ell = 1, \ldots, L$. Moreover, for any $f \in L^2(\mathbb{R})$ and $n, j \in \mathbb{N}$, (see [12], Lemma 3.1), we have

$$
(2.11) \qquad U_n^j = \sum_{r=(L+1)^j n}^{(L+1)^j(n+1)-1} U_r,
$$

and

$$
(2.12) \qquad \sum_{k\in\mathbb{Z}} |\langle f,\omega_{n,j,k}\rangle|^2 = \sum_{r=(L+1)^j n}^{(L+1)^j(n+1)-1} \sum_{k\in\mathbb{Z}} |\langle f,\omega_{r,0,k}\rangle|^2.
$$

By substituting $n=0$ in (2.11) and (2.12), we get

$$
(2.13) \qquad V_j = \sum_{r=0}^{(L+1)^j-1} U_r, \quad \text{and} \quad \sum_{k\in\mathbb{Z}} |\langle f,\varphi_{j,k}\rangle|^2 = \sum_{r=0}^{(L+1)^j-1} \sum_{k\in\mathbb{Z}} |\langle f,\omega_{r,0,k}\rangle|^2,
$$

for any $f \in L^2(\mathbb{R})$, respectively. Further, for $n = \ell$, $1 \leq \ell \leq L$, (2.11) and (2.12) yield

$$
(2.14) \qquad W_{j,\ell} = W_{0,\ell}^j = U_\ell^j = \sum_{r=(L+1)^j \ell}^{(L+1)^j(\ell+1)-1} U_r,
$$

and
(2.15)

$$
\sum_{k\in\mathbb{Z}} |\langle f,\psi_{\ell,j,k}\rangle|^2 = \sum_{k\in\mathbb{Z}} |\langle f,\omega_{\ell,j,k}\rangle|^2 = \sum_{r=(L+1)^j \ell}^{(L+1)^j(\ell+1)-1} \sum_{k\in\mathbb{Z}} |\langle f,\omega_{r,0,k}\rangle|^2,
$$

for any $f \in L^2(\mathbb{R})$, respectively. It is immediate from (2.15) that each wavelet space $W_{j,\ell}$, $\ell = 1,\ldots,L$, $j \geq 1$, can be further decomposed into $(L+1)^j$ subspaces U_r, $r \in [(L+1)^j \ell, (L+1)^j(\ell+1)-1]$. By choosing j to be fixed level $J > 0$, we have

$$
(2.16) \qquad L^2(\mathbb{R}) = \sum_{r=0}^{(L+1)^J-1} U_r + \sum_{\ell=1}^{L} \sum_{j\geq J} W_{j,\ell}.
$$

Theorem 2.2[12]. *For a given M-band tight wavelet frame $X(\Psi)$, the system*

$$\mathcal{F} = \left\{ \omega_{n,0,k} : 0 \leq n \leq (L+1)^J - 1, k \in \mathbb{Z} \right\}$$
$$\cup \left\{ \psi_{\ell,j,k} : \ell = 1, \ldots, L, j \geq J, k \in \mathbb{Z} \right\}$$

forms a tight frame for $L^2(\mathbb{R})$, where $\omega_n, n = 0, 1, \ldots,$ are the basic M-band framelet packets given by (2.9).

The concept of the basic M-band framelet packets enables us to construct various tight frames for $L^2(\mathbb{R})$ by choosing other $L^2(\mathbb{R})$ space decompositions. To do this, let Γ_J be the disjoint partition of a finite set of non-negative integers

(2.17) $$\Delta_J = \left\{ r \in \mathbb{N}_0 : 0 \leq r \leq (L+1)^J - 1 \right\}$$

into disjoint subsets of the form

$$I_{j,n} = \left\{ (L+1)^j n, \ldots, (L+1)^j (n+1) - 1 \right\}, \quad j, n \in \mathbb{N}_0,$$

i.e.,

(2.18) $$\Gamma_j = \left\{ I_{j,n} : \bigcup I_{j,n} = \Delta_J \right\}.$$

Theorem 2.3[12]. *Let Γ_J be a disjoint partition of Δ_J, where Γ_J and Δ_J are defined in (2.17) and (2.18), respectively. Then, the system*

$$\mathcal{F}_{\Gamma_J} = \left\{ \omega_{n,j,k} : I_{j,n} \in \Gamma_J, k \in \mathbb{Z} \right\} \cup \left\{ \psi_{\ell,j,k} : \ell = 1, \ldots, L, j \geq J, k \in \mathbb{Z} \right\}$$

generates a tight frame for $L^2(\mathbb{R})$, where $\omega_n, n = 0, 1, \ldots,$ are the basic M-band framelet packets given by (2.9).

3. MAIN RESULTS

Theorem 3.1. *Suppose $X(\Psi)$ is a M-band tight wavelet frame constructed via UEP in an MRA and $m_0, m_1, \ldots, m_L$ are the framelet symbols satisfying the UEP condition $\mathcal{M}(\xi)\mathcal{M}^*(\xi) = I_M$. Let $\omega_n, n = 0, 1, \ldots$, be as in equation (2.9). Assume that for $\alpha > 0$ there exists a positive constant C such that*

(3.1)
$$1 - |m_0(\xi)|^2 \leq C|\xi|^{2\alpha}, \quad \xi \in \mathbb{R}, \quad \text{and} \quad [\hat{\varphi}, \hat{\varphi}]_\alpha(\xi) \leq C, \quad \xi \in \mathbb{R}.$$

For any fixed $J > 0$, Γ_J is a disjoint partition of Δ_J, where Γ_J and Δ_J are defined in (2.17) and (2.18), respectively. Moreover, if $-\alpha < s < \alpha$, then the collection

$$\mathcal{F}^s_{\Gamma_J} = \left\{ M^{js}\omega_{n,j,k} : I_{j,n} \in \Gamma_J, k \in \mathbb{Z} \right\}$$

$$\cup \left\{ M^{js}\psi_{\ell,j,k} : \ell = 1, \ldots, L, j \geq J, k \in \mathbb{Z} \right\}$$

is a M-band framelet packet frame of $\mathbb{H}^s(\mathbb{R})$, i.e., there exist positive constants C_1, C_2 such that

$$C_1 \|f\|^2_{\mathbb{H}^s(\mathbb{R})} \leq \sum_{I_{j,n} \in \Gamma_J} \sum_{k \in \mathbb{Z}} M^{2js} |\langle f, \omega_{n,j,k} \rangle|^2$$

$$+ \sum_{\ell=1}^{L} \sum_{j=J}^{\infty} \sum_{k \in \mathbb{Z}} M^{2js} |\langle f, \psi_{\ell,j,k} \rangle|^2 \leq C_2 \|f\|^2_{\mathbb{H}^s(\mathbb{R})}$$

holds for all $f \in \mathbb{H}^s(\mathbb{R})$.

Proof. By Parseval's formula and the definition of bracket product (2.1) for $-\alpha < s < \alpha$, we have

$$\sum_{k\in\mathbb{Z}}|\langle f,\varphi_{J,k}\rangle|^2 = \frac{1}{2\pi}\sum_{k\in\mathbb{Z}}|\langle \hat{f},\hat{\varphi}_{J,k}\rangle|^2$$

$$= \frac{1}{2\pi}\int_T M^J \left|\left[\hat{f}(M^J\cdot),\hat{\varphi}\right](\xi)\right|^2 d\xi$$

$$\leq \frac{M^J}{2\pi}\int_T \left[\hat{f}(M^J\cdot),\hat{f}(M^J\cdot)\right]_{-s}(\xi)\,[\hat{\varphi},\hat{\varphi}]_s(\xi)\,d\xi$$

$$\leq \left\|[\hat{\varphi},\hat{\varphi}]_s\right\|_{L^\infty(\mathbb{R})} \frac{M^J}{2\pi}\int_T \left[\hat{f}(M^J\cdot),\hat{f}(M^J\cdot)\right]_{-s}(\xi)\,d\xi$$

$$\leq \left\|[\hat{\varphi},\hat{\varphi}]_\alpha\right\|_{L^\infty(\mathbb{R})} \frac{M^J}{2\pi}\int_\mathbb{R} \left|\hat{f}(M^J\xi)\right|^2 (1+|\xi|^2)^{-s}\,d\xi$$

$$\leq \frac{C}{2\pi}\int_\mathbb{R} \left|\hat{f}(\xi)\right|^2 (1+|M^{-J}\xi|^2)^{-s}\,d\xi$$

$$= \frac{C}{2\pi}\int_\mathbb{R} \left|\hat{f}(\xi)\right|^2 (1+|\xi|^2)^{-s} \left(\frac{1+|\xi|^2}{1+|M^{-J}\xi|^2}\right)^s d\xi$$

$$\leq C\max\{1,M^{2Js}\}\frac{1}{2\pi}\int_\mathbb{R}\left|\hat{f}(\xi)\right|^2(1+|\xi|^2)^{-s}\,d\xi$$

(3.2) $$= C\max\{1,M^{2Js}\}\|f\|_{\mathbb{H}^{-s}(\mathbb{R})}^2,$$

and in the last inequality, we used the fact that

$$1 \leq \frac{1+|\xi|^2}{1+|M^{-J}\xi|^2} \leq M^{2J}, \quad \xi\in\mathbb{R},\, J\in\mathbb{N}.$$

On the otherhand, we have

$$\sum_{I_{j,n}\in\Gamma_J}\sum_{k\in\mathbb{Z}} M^{-2js}|\langle f,\omega_{n,j,k}\rangle|^2 \le M^{2(J-1)|s|} \sum_{I_{j,n}\in\Gamma_J}\sum_{k\in\mathbb{Z}} |\langle f,\omega_{n,j,k}\rangle|^2$$

$$= M^{2(J-1)|s|} \sum_{I_{j,n}\in\Gamma_J} \sum_{r=(L+1)^j n}^{(L+1)^j(n+1)-1} \sum_{k\in\mathbb{Z}} |\langle f,\omega_{r,0,k}\rangle|^2$$

$$= M^{2(J-1)|s|} \sum_{n=0}^{(L+1)^J-1} \sum_{k\in\mathbb{Z}} |\langle f,\omega_{n,0,k}\rangle|^2$$

$$= M^{2(J-1)|s|} \sum_{k\in\mathbb{Z}} |\langle f,\varphi_{J,k}\rangle|^2$$

(3.3)
$$\le CM^{2(J-1)|s|} \max\{1, M^{2Js}\} \|f\|_{\mathbb{H}^{-s}(\mathbb{R})}^2.$$

Since $X(\Psi)$ is a M-band tight wavelet frame of $L^2(\mathbb{R})$, by [9, Proposition 2.1] and equation (3.1), we obtain

$$\sum_{\ell=1}^{L}\sum_{j=J}^{\infty}\sum_{k\in\mathbb{Z}} M^{-2js}|\langle f,\psi_{\ell,j,k}\rangle|^2 \le C\|B_{s,\ell,J}\|_{L^\infty(\mathbb{R})}\|f\|_{\mathbb{H}^{-s}(\mathbb{R})}^2,$$

where

$$B_{s,\ell,J} = \sum_{j=J}^{\infty} \frac{M^{-2js}(1+|\xi|^2)^s}{(1+|M^{-J}\xi|^2)^\alpha} \sum_{\ell=1}^{L} |m_\ell(M^{-j}\xi)|^2 \in L^\infty(\mathbb{R}).$$

Combining the inequalities (3.2) and (3.3), we obtain

$$\sum_{I_{j,n}\in\Gamma_J}\sum_{k\in\mathbb{Z}} M^{-2js}|\langle f,\omega_{n,j,k}\rangle|^2$$
$$+\sum_{\ell=1}^{L}\sum_{j=J}^{\infty}\sum_{k\in\mathbb{Z}} M^{-2js}|\langle f,\psi_{\ell,j,k}\rangle|^2 \le C'\|f\|_{\mathbb{H}^{-s}(\mathbb{R})}^2,$$

where $C' = C \left(\|B_{s,\ell,J}\|_{L^\infty(\mathbb{R})} + M^{2(J-1)|s|} \max\{1, M^{2Js}\} \right)$.

By duality argument as in the proof of [9, Theorem 1.2], we can obtain

$$\frac{1}{C'}\|f\|^2_{\mathbb{H}^s(\mathbb{R})} \leq \sum_{I_{j,n}\in\Gamma_J}\sum_{k\in\mathbb{Z}} M^{2js}|\langle f, \omega_{n,j,k}\rangle|^2$$

$$+ \sum_{\ell=1}^{L}\sum_{j=J}^{\infty}\sum_{k\in\mathbb{Z}} M^{2js}|\langle f, \psi_{\ell,j,k}\rangle|^2 \leq C'\|f\|^2_{\mathbb{H}^s(\mathbb{R})},$$

for all $f \in \mathbb{H}^s(\mathbb{R}), -\alpha < s < \alpha$. This completes the proof.

REFERENCES

[1] Bi, N., Dai, X. and Sun, Q., *Construction of compactly supported M-band wavelets*, Appl. Comput. Harmon. Anal., **6** (1999), 113-131.

[2] Chui, C. K. and Li, C., *Non-orthogonal wavelet packets*, SIAM J. Math. Anal., **24** (1993), 712-738.

[3] Coifman, R. R., Meyer, Y., Quake, S. and Wickerhauser, M. V., *Signal processing and compression with wavelet packets*, Technical Report, Yale University, 1990.

[4] Daubechies, I., Han, B., Ron, A. and Shen, Z., *Framelets: MRA-based constructions of wavelet frames*, Appl. Comput. Harmon. Anal., **14** (2003), 1-46.

[5] Han, J. C. and Cheng, Z., *The construction of M-band tight wavelet frames*, In Proceedings of the Third Intemational Conference on Machine Learning and Cybemetics, Shanghai. 2004, pp. 26-29.

[6] Jiankang, Z., Zheng, B. and Licheng, J., *Theory of orthonormal M-band wavelet packets*, J. Elect., **15**(3) (1998), 193-198.

[7] Lin, T., Xu, S., Shi, Q. and Hao, P., *An algebraic construction of orthonormal M-band wavelets with perfect reconstruction*, Appl. Comput. Harmon. Anal., **17** (2006), 717-730.

[8] Petukhov, A. P., *Explicit construction of framelets*, Appl. Comput. Harmon. Anal., **11** (2001), 313-327.

[9] Ron, A. and Shen, Z., *Affine systems in $L^2(\mathbb{R}^d)$: the analysis of the analysis operator*, J. Funct. Anal., **148** (1997), 408-447.

[10] Shah, F. A., *Construction of wavelet packets on p-adic field*, Int. J. Wavelets, Multiresolut. Inf. Process., **7**(5) (2009), 553-565.

[11] Shah, F. A. and Debnath, L., *p-Wavelet frame packets on a half-line using the Walsh-Fourier transform*, Integ. Transf. Spec. Funct., **22**(12) (2011), 907-917.

[12] Shah, F. A. and Debnath, L., *Explicit construction of M-band tight framelet packets*, Analysis, **32** (2012), 281-294.

[13] Steffen, P., Heller, P. N., Gopinath, R. A. and Burrus, C. S., *Theory of regular M-band wavelet bases*, IEEE Trans. Sig. Proces., **41**(12) (1993), 3497-3510.

*Department of Mathematics, University of Kashmir, South Campus, Anantnag-192 101, Jammu and Kashmir, India.
E-mail: fashah79@gmail.com

**Department of Mathematics, Jamia Millia Islamia, New Delhi-110025, India.
E-mail: khalil_ahmad49@yahoo.com

Differential Geometry, Functional Analysis and Applications
Editors: Mohammad Hasan Shahid, Sharfuddin Ahmad *et al.*
Copyright © 2015, Narosa Publishing House, New Delhi

APPROXIMATION OF BOUND FOR THE CLASS OF POLYNOMIALS VANISHING INSIDE THE DISK

ARTY AHUJA AND K.K. DEWAN

ABSTRACT. Let $p(z)$ be a polynomial of degree n having all its zeros in $|z| \leq k$, $k > 1$, then Aziz [Bull. Austral. Math. Soc., 35 (1987)] proved that

$$M(p, R) \geq \left(\frac{R+k}{1+k}\right)^n M(p, 1) \text{ for } R \geq k^2.$$

The above inequality is valid for $k > 1$ and $R \geq k^2$. In this paper, we have obtained a similar type of result for lacunary type of polynomials $p(z) = a_n z^n + \sum_{j=\mu}^{n} a_{n-j} z^{n-j}$, $1 \leq \mu \leq n$ having all its zeros in $|z| \leq k$, $k > 1$ where $R < k^2$. Our result also improves upon the result due to Dewan, Singh and Yadav [Southeast Asian Bull. Math., 27 (2) (2003)], Dewan and Upadhyaye [Ind. J. Pure Appl. Math., 30 (2007)].

1. INTRODUCTION

Let $p(z)$ be a polynomial of degree n and let $M(p, \delta) = \max_{|z|=\delta} |p(z)|$ ($0 \leq \delta < \infty$). Then applying the Maximum Modulus Principle to the polynomial

$$p^*(z) = z^n \, \overline{p\left(\frac{1}{\bar{z}}\right)},$$

we see that

(1.1) $\quad M(p,r) = r^n M(p^*, r^{-1}) \geq r^n M(p^*, 1) = r^n M(p, 1) \quad (0 \leq r < 1),$

where equality holds if and only if $p(z) = cz^n$, $c \neq 0$.

For polynomials not vanishing in $|z| < 1$, Rivlin [7] obtained a stronger inequality and proved that if $p(z) = \sum_{\nu=0}^{n} a_\nu z^\nu$ is a polynomial of degree

---

2000 *Mathematics Subject Classification.* 30A10, 30C10, 30C15.
Key words and phrases. Maximum Modulus, Polynomials, Zeros.

n which does not vanish in $|z| < 1$, then

(1.2) $$M(p,r) \geq \left(\frac{r+1}{2}\right)^n M(p,1) \text{ for } 0 \leq r < 1.$$

Here equality is attained if $p(z) = \alpha(z-\beta)^n$, $|\beta| = 1$.

Aziz [1] obtained the following result for the class of polynomials having all its zeros in $|z| \leq k$, $k > 1$. In fact, he proved

Theorem A. *If* $p(z) = \sum_{\nu=0}^{n} a_\nu z^\nu$ *is a polynomial of degree n having all its zeros in $|z| \leq k$, $k > 1$, then*

(1.3) $$M(p,R) \geq \left(\frac{R+k}{1+k}\right)^n M(p,1) \text{ for } R \geq k^2.$$

Inequality (1.3) is valid for $k > 1$ and $R \geq k^2$. Jain [5] obtained similar type of inequality for $k > 1$ and $R < k^2$ by proving the following theorem.

Theorem B. *Let* $p(z) = \sum_{\nu=0}^{n} a_\nu z^\nu$ *be a polynomial of degree n having all its zeros in $|z| \leq k$, $k > 1$. Then for $k < R < k^2$*

(1.4) $$M(p,R) \geq R^s \left(\frac{R+k}{1+k}\right) M(p,1) \text{ for } s < n,$$

where s is the order of a possible zero of $p(z)$ at $z = 0$.

Dewan and Upadhyaye [4] considered the class of polynomials $p(z) = a_n z^n + \sum_{j=\mu}^{n} a_{n-j} z^{n-j}$, $1 \leq \mu \leq n$, of degree n having all its zeros in $|z| \leq k$, $k > 1$ and extended Theorem B to lacunary polynomial in the following manner.

Theorem C. *Let* $p(z) = a_n z^n + \sum_{j=\mu}^{n} a_{n-j} z^{n-j}$, $1 \leq \mu \leq n$ *be a polynomial of degree n having all its zeros in $|z| \leq k$, $k > 1$. Then for $k < R < k^2$*

(1.5) $$M(p,R) \geq R^s \left(\frac{R^\mu + k^\mu}{R^{\mu-1} + k^\mu}\right)^n M(p,1),$$

where s is the order of a possible zero of $p(z)$ at $z = 0$ and $0 \leq s \leq n - \mu$.

By involving coefficients Barchand [2] generalized Theorem C and proved that

134

Theorem D. Let $p(z) = a_n z^n + \sum_{j=\mu}^{n} a_{n-j} z^{n-j}$, $1 \leq \mu \leq n$, be a polynomial of degree n having all its zeros in $|z| \leq k$, $k > 1$. Then for $k < R < k^2$

(1.6) $$\max_{|z|=R} |p(z)|$$
$$\geq R^s \frac{\begin{pmatrix} (R^{n-s-1}k^{2\mu} + R^{\mu-s+\mu}k^{\mu-1})(n-s)|a_n| \\ +\mu |a_{n-\mu}|(R^{n-s+\mu-1} + R^{n-s}k^{\mu-1}) \end{pmatrix}}{\begin{pmatrix} (R^{n-s-1}k^{2\mu} + R^{\mu}k^{\mu-1})(n-s)|a_n| \\ +\mu |a_{n-\mu}|(R^{\mu-1} + R^{n-s}k^{\mu-1}) \end{pmatrix}} \max_{|z|=1} |p(z)|,$$

where s is the order of a possible zero of $p(z)$ at origin with $s \leq n - \mu$.

If we involve $m = \min_{|z|=k} |p(z)|$ also, then we are able to improve upon Theorems C and D for the class of polynomials $p(z) = a_n z^n + \sum_{j=\mu}^{n} a_{n-j} z^{n-j}$, $1 \leq \mu \leq n$, having all its zeros in $|z| \leq k$, $k > 1$. More precisely, we prove that

Theorem. Let $p(z) = a_n z^n + \sum_{j=\mu}^{n} a_{n-j} z^{n-j}$, $1 \leq \mu \leq n$, be a polynomial of degree n having all its zeros in $|z| \leq k$, $k > 1$. Then for $k < R < k^2$

(1.7) $$\max_{|z|=R} |p(z)|$$
$$\geq R^s \frac{\begin{pmatrix} (R^{n-s-1}k^{2\mu} + R^{n-s+\mu}k^{\mu-1})(n-s)|a_n| \\ +\mu |a_{n-\mu}|(R^{n-s+\mu-1} + R^{n-s}k^{\mu-1}) \end{pmatrix}}{\begin{pmatrix} (R^{n-s-1}k^{2\mu} + R^{\mu}k^{\mu-1})(n-s)|a_n| \\ +\mu |a_{n-\mu}|(R^{\mu-1} + R^{n-s}k^{\mu-1}) \end{pmatrix}} \max_{|z|=1} |p(z)|$$
$$+ \frac{R^{s+\mu-1}}{k^s} \frac{\begin{pmatrix} (R^{n-s} - 1) \\ \times ((n-s)|a_n|Rk^{\mu-1} + \mu |a_{n-\mu}|) \end{pmatrix}}{\begin{pmatrix} (k^{2\mu} + R^{n-s-1} + k^{\mu-1}R^{\mu})(n-s)|a_n| \\ +\mu |a_{n-\mu}|(k^{\mu-1} + R^{n-s} + R^{\mu-1}) \end{pmatrix}} \min_{|z|=k} |p(z)|,$$

where s is the order of a possible zero of $p(z)$ at origin with $s \leq n - \mu$.

In particular, for $\mu = 1$, we get

Corollary 1. *Let $p(z) = \sum_{j=0}^{n} a_j z^j$ be a polynomial of degree n having all its zeros in $|z| \leq k$, $k > 1$, then for $k < R < k^2$*

$$(1.8) \quad \max_{|z|=R} |p(z)| \geq R^s \frac{\begin{pmatrix} \{(R^{n-s-1}k^2 + R^{n-s+1}) \\ \times (n-s)|a_n| + 2|a_{n-1}|R^{n-s}\} \end{pmatrix}}{\begin{pmatrix} (R^{n-s-1}k^2 + R)(n-s)|a_n| \\ + |a_{n-1}|(1 + R^{n-s}) \end{pmatrix}} \max_{|z|=1} |p(z)|$$

$$+ \frac{R^s}{k^s} \frac{(R^{n-s} - 1)\{(n-s)|a_n|R + |a_{n-1}|\}}{\begin{pmatrix} (k^2 R^{n-s-1} + R)(n-s)|a_n| \\ + |a_{n-1}|(R^{n-s} + 1) \end{pmatrix}} \min_{|z|=k} |p(z)|,$$

where s is the order of a possible zero of $p(z)$ at origin with $s \leq n - 1$.

It can be easily seen that Corollary 1 is an improvement of Theorem B.

2. Lemmas

Lemma 1. *If $p(z) = a_0 + \sum_{\nu=\mu}^{n} a_\nu z^\nu$, $1 \leq \mu \leq n$, is a polynomial of degree n having all its zeros in $|z| \geq k$, $k \geq 1$, then*

(2.1)
$$\max_{|z|=1} |p'(z)| \leq n \left(\frac{n|a_0| + \mu|a_\mu|k^{\mu+1}}{n|a_0|(1 + k^{\mu+1}) + \mu|a_\mu|(k^{\mu+1} + k^{2\mu})} \right) \max_{|z|=1} |p(z)|$$
$$- \frac{n}{k^n} \left(\frac{n|a_0|k^{\mu+1} + \mu|a_\mu|k^{2\mu}}{n|a_0|(1 + k^{\mu+1}) + \mu|a_\mu|(k^{\mu+1} + k^{2\mu})} \right) \min_{|z|=k} |p(z)|.$$

The above lemma is due to Dewan, Singh and Yadav [3].

Lemma 2. *If $p(z) = a_0 + \sum_{\nu=\mu}^{n} a_\nu z^\nu$, $1 \leq \mu \leq n$, is a polynomial of degree n having all its zeros in $|z| \geq k$, $k \geq 1$, then for $r \leq k \leq R$*

(2.2) $$\max_{|z|=R}|p(z)| \leq \frac{\begin{pmatrix} n|a_0|(R^n + k^{\mu+1}r^{n-\mu-1})r^\mu \\ +\mu|a_\mu|r^{\mu-1}k^{\mu+1}(R^n + k^{\mu-1}r^{n-\mu+1}) \end{pmatrix}}{\begin{pmatrix} n|a_0|(r^{\mu+1} + k^{\mu+1})r^{n-1} \\ +\mu|a_\mu|r^n k^{\mu+1}(r^{\mu-1} + k^{\mu-1}) \end{pmatrix}} \max_{|z|=r}|p(z)|$$
$$- \frac{1}{k^n}\frac{r^{n-1}(R^n - r^n)(n|a_0|k^{\mu+1} + \mu|a_\mu|k^{2\mu}r)}{\begin{pmatrix} n|a_0|(r^{\mu+1} + k^{\mu+1})r^{n-1} \\ +\mu|a_\mu|r^n k^{\mu+1}(r^{\mu-1} + k^{\mu-1}) \end{pmatrix}} \min_{|z|=k}|p(z)|.$$

Proof of Lemma 2. Let $0 \leq r \leq k$. Since $p(z)$ is a polynomial of degree n having no zero in $|z| < k$, $k \geq 1$, the polynomial $T(z) = p(rz)$ has no zero in $|z| < \frac{k}{r}$, $\frac{k}{r} \geq 1$, therefore, applying Lemma 1 to $T(z)$, we get

$$\max_{|z|=1}|T'(z)|$$
$$\leq n \frac{n|a_0| + \mu|r^\mu a_\mu|\frac{k^{\mu+1}}{r^{\mu+1}}}{n|a_0|\left(1 + \frac{k^{\mu+1}}{r^{\mu+1}}\right) + \mu|r^\mu a_\mu|\left(\frac{k^{\mu+1}}{r^{\mu+1}} + \frac{k^{2\mu}}{r^{2\mu}}\right)} \max_{|z|=1}|T(z)|$$
$$- \frac{nr^n}{k^n}\frac{n|a_0|\frac{k^{\mu+1}}{r^{\mu+1}} + \mu|r^\mu a_\mu|\frac{k^{2\mu}}{r^{2\mu}}}{n|a_0|\left(1 + \frac{k^{\mu+1}}{r^{\mu+1}}\right) + \mu|r^\mu a_\mu|\left(\frac{k^{\mu+1}}{r^{\mu+1}} + \frac{k^{2\mu}}{r^{2\mu}}\right)} \min_{|z|=\frac{k}{r}}|T(z)|.$$

Replacing $T(z)$ by $p(rz)$, we get

(2.3) $$\max_{|z|=r}|p'(z)| \leq n\frac{n|a_0|r^\mu + \mu|a_\mu|r^{\mu-1}k^{\mu+1}}{\begin{pmatrix} n|a_0|(r^{\mu+1} + k^{\mu+1}) \\ +\mu|a_\mu|(k^{\mu+1}r^\mu + k^{2\mu}r) \end{pmatrix}} \max_{|z|=r}|p(z)|$$

$$-\frac{nr^{n-1}}{k^n}\frac{n|a_0|k^{\mu+1}+\mu|a_\mu|rk^{2\mu}}{\begin{pmatrix}n|a_0|(r^{\mu+1}+k^{\mu+1})\\+\mu|a_\mu|(k^{\mu+1}r^\mu+k^{2\mu}r)\end{pmatrix}}\min_{|z|=k}|p(z)|.$$

Since $p'(z)$ is a polynomial of degree at most $(n-1)$, then by Maximum Modulus Principle [6, p. 158, Problem III, 269], we have

$$\frac{M(p',t)}{t^{n-1}} \leq \frac{M(p',R)}{r^{n-1}} \text{ for } t \geq r.$$

The above inequality in conjunction with (2.3), yields

$$\max_{|z|=t}|p'(z)| \leq \frac{nt^{n-1}}{r^{n-1}}\left(\frac{n|a_0|r^\mu+\mu|a_\mu|r^{\mu-1}k^{\mu+1}}{\begin{pmatrix}n|a_0|(r^{\mu+1}+k^{\mu+1})\\+\mu|a_\mu|(k^{\mu+1}r^\mu+k^{2\mu}r)\end{pmatrix}}\max_{|z|=r}|p(z)|\right.$$

$$\left.-\frac{r^{n-1}}{k^n}\frac{n|a_0|k^{\mu+1}+\mu|a_\mu|rk^{2\mu}}{\begin{pmatrix}n|a_0|(r^{\mu+1}+k^{\mu+1})\\+\mu|a_\mu|(k^{\mu+1}r^\mu+k^{2\mu}r)\end{pmatrix}}\min_{|z|=k}|p(z)|\right).$$

Now, for $0 \leq \theta < 2\pi$, we have

$$|p(Re^{i\theta})-p(re^{i\theta})|$$

$$\leq \int_r^R |p'(te^{ie})|dt$$

$$\leq \frac{n}{r^{n-1}}\left(\frac{n|a_0|r^\mu+\mu|a_\mu|r^{\mu-1}k^{\mu+1}}{\begin{pmatrix}n|a_0|(r^{\mu+1}+k^{\mu+1})\\+\mu|a_\mu|(k^{\mu+1}r^\mu+k^{2\mu}r)\end{pmatrix}}\max_{|z|=r}|p(z)|\right.$$

$$\left.-\frac{r^{n-1}}{k^n}\frac{n|a_0|k^{\mu+1}+\mu|a_\mu|rk^{2\mu}}{\begin{pmatrix}n|a_0|(r^{\mu+1}+k^{\mu+1})\\+\mu|a_\mu|(k^{\mu+1}r^\mu+k^{2\mu}r)\end{pmatrix}}\min_{|z|=k}|p(z)|\right)\int_r^R t^{n-1}dt$$

$$= \frac{R^n - r^n}{r^{n-1}} \left(\frac{n|a_0|r^\mu + \mu|a_\mu|r^{\mu-1}k^{\mu+1}}{\begin{pmatrix} n|a_0|(r^{\mu+1} + k^{\mu+1}) \\ +\mu|a_\mu|(k^{\mu+1}r^\mu + k^{2\mu}r) \end{pmatrix}} \max_{|z|=r}|p(z)| \right.$$

$$\left. - \frac{r^{n-1}}{k^n} \frac{n|a_0|k^{\mu+1} + \mu|a_\mu|rk^{2\mu}}{\begin{pmatrix} n|a_0|(r^{\mu+1} + k^{\mu+1}) \\ +\mu|a_\mu|(k^{\mu+1}r^\mu + k^{2\mu}r) \end{pmatrix}} \min_{|z|=k}|p(z)| \right).$$

$$M(p,R) \leq \left[1 + \left(\frac{R^n - r^n}{r^{n-1}}\right) \frac{n|a_0|r^\mu + \mu|a_\mu|r^{\mu-1}k^{\mu+1}}{\begin{pmatrix} n|a_0|(r^{\mu+1} + k^{\mu+1}) \\ +\mu|a_\mu|(k^{\mu+1}r^\mu + k^{2\mu}r) \end{pmatrix}} \max_{|z|=r}|p(z)| \right.$$

$$\left. - \left(\frac{R^n - r^n}{k^n}\right) \frac{n|a_0|k^{\mu+1} + \mu|a_\mu|rk^{2\mu}}{\begin{pmatrix} n|a_0|(r^{\mu+1} + k^{\mu+1}) \\ +\mu|a_\mu|(k^{\mu+1}r^\mu + k^{2\mu}r) \end{pmatrix}} \min_{|z|=k}|p(z)| \right),$$

from which the result follows. □

3. Proof of Theorem

Proof of Theorem. Since the polynomial $p(z)$ has all its zeros in $|z| \leq k$, $k > 1$ with s-fold zeros at the origin, therefore, the polynomial $q(z) = z^n \overline{p\left(\frac{1}{\bar{z}}\right)}$ has all its zeros in $|z| \leq \frac{1}{k}$, $\frac{1}{k} < 1$ and is of degree $(n-s)$.

On applying Lemma 2 to the polynomial $q(z)$ with $R = 1$, we obtain for $\frac{1}{k^2} < r < \frac{1}{k}$

$$\max_{|z|=r}|q(z)| \geq \frac{\begin{pmatrix} r^{n-s-1}\left(r^{\mu+1} + \frac{1}{k^{\mu+1}}\right)(n-s)|a_n| \\ +\mu|a_{n-\mu}|r^{n-s}\frac{1}{k^{\mu+1}}\left(r^{\mu+1} + \frac{1}{k^{\mu-1}}\right) \end{pmatrix}}{\begin{pmatrix} \left(1 + \frac{r^{n-s-\mu-1}}{k^{\mu+1}}\right)r^\mu(n-s)|a_n| \\ +\mu|a_{n-\mu}|\frac{r^{\mu-1}}{k^{\mu+1}}\left(1 + \frac{r^{n-s-\mu-1}}{k^{\mu-1}}\right) \end{pmatrix}} \max_{|z|=1}|q(z)|$$

$$+ \frac{r^{n-s-1}(1-r^{n-s})\left\{(n-s)|a_n|\frac{1}{k^{\mu+1}} + \mu|a_{n-\mu}|\frac{r}{k^{2\mu}}\right\}}{\frac{1}{k^{n-s}}\left(\left[\left(1+\frac{r^{n-s-\mu-1}}{k^{\mu+1}}\right)r^\mu(n-s)|a_n| \atop +\mu|a_{n-\mu}|\frac{r^{\mu-1}}{k^{\mu+1}}\left(1+\frac{r^{n-s-\mu+1}}{k^{\mu-1}}\right)\right]\right)} \min_{|z|=\frac{1}{k}}|q(z)|$$

which implies

$$\max_{|z|=r}\left|z^n\overline{p\left(\frac{1}{\bar{z}}\right)}\right|$$

$$\geq \frac{r^{n-s-1}\left(\left\{\left(r^{\mu+1}+\frac{1}{k^{\mu+1}}\right)(n-s)|a_n| \atop +\mu|a_{n-\mu}|\frac{r}{k^{\mu+1}}\left(r^{\mu-1}+\frac{1}{k^{\mu-1}}\right)\right\}\right)}{\left(\left(1+\frac{r^{n-s-\mu-1}}{k^{\mu+1}}\right)r^\mu(n-s)|a_n| \atop +\mu|a_{n-\mu}|\frac{r^{\mu-1}}{k^{\mu+1}}\left(1+\frac{r^{n-s-\mu+1}}{k^{\mu-1}}\right)\right)} \max_{|z|=1}|p(z)|$$

$$+ \frac{\left(r^{n-s-1}k^{n-s}(1-r^{n-s}) \atop \times\left\{(n-s)|a_n|\frac{1}{k^{\mu+1}} + \mu|a_{n-\mu}|\frac{r}{k^{2\mu}}\right\}\right)}{\left(\left(1+\frac{r^{n-s-\mu-1}}{k^{\mu+1}}\right)r^\mu(n-s)|a_n| \atop +\mu|a_{n-\mu}|\frac{r^{\mu-1}}{k^{\mu+1}}\left(1+\frac{r^{n-s-\mu+1}}{k^{\mu-1}}\right)\right)} \min_{|z|=\frac{1}{k}}\left|z^n q\left(\frac{1}{z}\right)\right|$$

This is equivalent to

$$\max_{|z|=\frac{1}{r}}|p(z)| \geq \frac{\left(r^{-s-1}\left\{\left(r^{\mu+1}+\frac{1}{k^{\mu+1}}\right)(n-s)|a_n| \atop +\mu|a_{n-\mu}|\left(\frac{r^\mu}{k^{\mu+1}}+\frac{r}{k^{2\mu}}\right)\right\}\right)}{\left(\left(r^\mu+\frac{r^{n-s-1}}{k^{\mu+1}}\right)(n-s)|a_n| \atop +\mu|a_{n-\mu}|\left(\frac{r^{\mu-1}}{k^{\mu+1}}+\frac{r^{n-s}}{k^{2\mu}}\right)\right)} \max_{|z|=1}|p(z)|$$

$$+ \frac{\begin{pmatrix} r^{-s-1}k^{-s}(1-r^{n-s}) \\ \times \left\{ (n-s)|a_n|\frac{1}{k^{\mu+1}} + \mu|a_{n-\mu}|\frac{r}{k^{2\mu}} \right\} \end{pmatrix}}{\begin{pmatrix} \left(r^\mu + \frac{r^{n-s-1}}{k^{\mu+1}} \right)(n-s)|a_n| \\ +\mu|a_{n-\mu}| \left(\frac{r^{\mu-1}}{k^{\mu+1}} + \frac{r^{n-s}}{k^{2\mu}} \right) \end{pmatrix}} \min_{|z|=k} |p(z)|.$$

Now replacing r by $\frac{1}{R}$ in above inequality so that $\frac{1}{k^2} < \frac{1}{R} < \frac{1}{k}$ or $k < R < k^2$, we get

$$\max_{|z|=R} |p(z)|$$

$$\geq \frac{R^{s+1} \begin{pmatrix} \left\{ \left(\frac{1}{R^{\mu+1}} + \frac{1}{k^{\mu+1}} \right)(n-s)|a_n| \\ +\mu|a_{n-\mu}| \left(\frac{1}{R^\mu k^{\mu+1}} + \frac{1}{Rk^{2\mu}} \right) \right\} \end{pmatrix}}{\begin{pmatrix} \left(\frac{1}{R^\mu} + \frac{1}{R^{n-s-1}k^{\mu+1}} \right)(n-s)|a_n| \\ +\mu|a_{n-\mu}| \left(\frac{1}{R^{\mu-1}k^{\mu+1}} + \frac{1}{R^{n-s}k^{2\mu}} \right) \end{pmatrix}} \max_{|z|=1} |p(z)|$$

$$+ \frac{R^{s+1}}{k^s} \frac{\left(1 - \frac{1}{R^{n-s}} \right) \left\{ (n-s)|a_n|\frac{1}{k^{\mu+1}} + \mu|a_{n-\mu}|\frac{1}{k^{2\mu}R} \right\}}{\begin{pmatrix} \left(\frac{1}{R^\mu} + \frac{1}{R^{n-s-1}k^{\mu+1}} \right)(n-s)|a_n| \\ +\mu|a_{n-\mu}| \left(\frac{1}{R^{\mu-1}k^{\mu+1}} + \frac{1}{R^{n-s}k^{2\mu}} \right) \end{pmatrix}} \min_{|z|=k} |p(z)|$$

which on simplification reduces to

$$\max_{|z|=R} |p(z)| \geq R^s \left\{ \frac{\begin{array}{l}(R^{n-s-1}k^{2\mu} + R^{n-s+\mu}k^{\mu-1})(n-s)|a_n| \\ +\mu|a_{n-\mu}|(R^{n-s+\mu-1} + R^{n-s}k^{\mu-1})\end{array}}{\begin{array}{l}(R^{n-s-1}k^{2\mu} + R^\mu k^{\mu-1})(n-s)|a_n| \\ +\mu|a_{n-\mu}|(R^{\mu-1} + R^{n-s}k^{\mu-1})\end{array}} \right\} \max_{|z|=1} |p(z)|$$

$$+ \frac{R^{s+\mu-1}}{k^s} \frac{(R^{ns}-1)\{(n-s)|a_n|Rk^{\mu-1}+\mu|a_{n-\mu}|\}}{\left\{ \begin{array}{c} (R^{n-s-1}k^{2\mu}+R^\mu k^{\mu-1})(n-s)|a_n| \\ +\mu|a_{n-\mu}|(R^{\mu-1}+R^{n-s}k^{\mu-1}) \end{array} \right\}} \min_{|z|=k}|p(z)|.$$

This completes the proof of Theorem. □

References

[1] Aziz, A., *Growth of polynomials whose zeros are within or outside a circle*, Bull. Austral. Math. Soc., **35** (1987), 247–250.

[2] Barchand, C., *Growth of polynomials*, Ph.D. Thesis submitted to Jamia Millia Islamia, New Delhi, (2002).

[3] Dewan, K. K., Singh, H. and Yadav, R. S., *Inequalities concerning polynomials having zeros in closed exterior or closed interior of a circle*, Southeast Asian Bulletin of Math., **27** (2) (2003), 591–597.

[4] Dewan, K. K. and Upadhyaye, C., *On maximum modulus of polynomials*, Indian J. Pure and Appl. Math., **38** (4) (2007), 325–330.

[5] Jain, V. K., *On polynomials having zeros in closed exterior or interior of a circle*, Indian J. Pure Appl. Math., **30** (1999), 153–159.

[6] Polya, G. and Szego, G., *Aufgaben and Lehrsatze aus der Analysis*, Springer-Verlag, Berlin, (1925).

[7] Rivlin, T. J., *On the maximum modulus of polynomials*, Amer. Math. Monthly, **67** (1960), 251–253.

Department of Mathematics,
Faculty of Natural Sciences
Jamia Millia Islamia (Central University),
New Delhi 110025 (India)
E-mail: aarty_ahuja@yahoo.com

Differential Geometry, Functional Analysis and Applications
Editors: Mohammad Hasan Shahid, Sharfuddin Ahmad *et al.*
Copyright © 2015, Narosa Publishing House, New Delhi

STRONG AND Δ-CONVERGENCE OF KHAN ET. AL. ITERATIVE PROCEDURE IN CAT(0) SPACES

MADHU AGGARWAL[1] AND RENU CHUGH[2]

ABSTRACT. In this paper, we prove some strong and Δ- convergence theorems of Khan et. al. iterative procedure in CAT(0) spaces which converges at a rate similar to that of Picard and Agarwal et. al. but faster than Mann, Ishikawa-type and one studied by Yao and Chen[22]. The results obtained are extension of some recent results of Khan and Abbas[11] to the case of two mappings.

1. INTRODUCTION AND PRELIMINARIES

Fixed point theory in CAT(0) spaces was first studied by kirk(see [14] and [15]). He showed that every nonexpansive (singlevalued) mapping defined on a bounded closed convex subset of a complete CAT(0) space always has a fixed point. In 2008, Kirk and Panyanak[13] generalized Lims[17] concept of Δ-convergence in CAT(0) spaces to prove the CAT(0) space analogs of some Banach space results which involve weak convergence, and Dhompongsa and Panyanak[7] obtained Δ-convergence theorems for the Picard, Mann and Ishikawa iterative procedures in the CAT(0) space setting. Afterwards, Panyanak and Laokul[19], Beg and Abbas[2], Shahzad[21], Chaoha and Phon-on[5]and Laokul and Panyanak[16] continued to work in this direction and obtained some results using Mann and Ishikawa iterative procedures involving one mapping. In 2011, Khan and Abbas[11] obtained strong and Δ-convergence theorems for Agarwal et. al. iterative procedure which is both faster than and independent of the Ishikawa iterative procedure. They also obtained some convergence results for two mappings using the Ishikawa-type iterative procedure. In this paper, we prove some

2000 *Mathematics Subject Classification.* 47H09,47H10, 54H25.
Key words and phrases. CAT(0)spaces, Δ-convergence, strong convergence, nonexpansive mappings, common fixed points, Iterative procedures.

strong and Δ-convergence results for approximating common fixed points of two nonexpansive self mappings in CAT(0) spaces using Khan et. al. iterative procedure.

Now, we recall some well known concepts and results.

Throughout this paper, N denotes the set of all positive integers and R denotes the set of all real numbers.

Let (X, d) be a metric space. A geodesic path joining $x \in X$ to $y \in X$ (or, more briefly, a geodesic from x to y) is a map c from a closed interval $[0, l] \subset R$ to X such that $c(0) = x, c(l) = y$ and $d(c(t), c(\acute{t})) = |t - \acute{t}|$ for all $t, \acute{t} \in [0, l]$. In particular, c is an isometry and $d(x, y) = l$. Usually, the image $c([0, l])$ of c is called a geodesic (or metric) segment joining x and y. A geodesic segment joining x and y is not necessarily unique in general. In particular, in the case when the geodesic segment joining x and y is unique, we use $[x, y]$ to denote the unique geodesic segment joining x and y.

The space (X, d) is said to be a geodesic space, if every two points of X are joined by a geodesic, and X is said to be uniquely geodesic space, if there is exactly one geodesic joining x and y, for each $x, y \in X$. A subset $Y \subseteq X$ is said to be convex, if Y includes every geodesic segment joining any two of its points.

A geodesic triangle $\Delta(x_1, x_2, x_3)$ in a geodesic metric space (X, d) consists of three points $x_1, x_2, x_3 \in X$ (the vertices of Δ) and a geodesic segment between each pair of vertices (the edges of Δ). A comparison triangle for the geodesic triangle $\Delta(x_1, x_2, x_3)$ in (X, d) is a triangle $\bar{\Delta}(x_1, x_2, x_3) := \Delta(\bar{x}_1, \bar{x}_2, \bar{x}_3)$ in the Euclidean plane E^2 such that $d_{E^2}(\bar{x}_i, \bar{x}_j) = d(x_i, x_j)$ for $i, j \in \{1, 2, 3\}$. The point $\bar{p} \in [\bar{x}, \bar{y}]$ is called a comparison point in $\bar{\Delta}$ for $p \in [x, y]$ if $d(x, p) = d_{E^2}(\bar{x}, \bar{p})$.

A geodesic space is said to be a CAT(0) space, if all geodesic triangles satisfy the following comparison axiom.

CAT(0): Let Δ be a geodesic triangle in X and Let $\bar{\Delta}$ be comparison triangle for Δ. Then Δ is said to satisfy CAT(0) inequality if for all $x, y \in \Delta$ and all comparison points $\bar{x}, \bar{y} \in \bar{\Delta}$, $d(x, y) \leq d_{E^2}(\bar{x}, \bar{y})$.

If x, y_1, y_2 are points of a CAT(0) space and if y_0 is the midpoint of the segment $[y_1, y_2]$ then the CAT(0) inequality implies

$$(CN) \qquad d(x, y_0)^2 \leq \frac{1}{2} d(x, y_1)^2 + \frac{1}{2} d(x, y_2)^2 - \frac{1}{4} d(y_1, y_2)^2.$$

This is the (CN) inequality of Bruhat and Tits[4]. In fact, (c.f. [3], p. 163), a geodesic space is a CAT(0) space if and only if it satisfies (CN) inequality.

Remark 1.1. *For $\kappa < 0$, a $CAT(\kappa)$ space is defined in terms of comparison triangles in the hyperbolic plane (see [3] for details). Here, for sake of simplicity, we omit definition, since it is known (see[3], page 165) that any $CAT(\kappa_1)$ space is also $CAT(\kappa_2)$ space for any pair (κ_1, κ_2) with $\kappa_2 \geq \kappa_1$. This means that the results in CAT(0) spaces can be applied to $CAT(\kappa)$ spaces with $\kappa \leq 0$.*

Lemma 1.1 ([7]). *Let (X, d) be a CAT(0) space. Then*

(i) *(X, d) is uniquely geodesic.*

(ii) *Let $p, x, y \in X$ and $\alpha \in [0, 1]$. Let m_1, m_2 denote, respectively, $[p, x], [p, y]$ satisfying $d(p, m_1) = \alpha d(p, x), d(p, m_2) = \alpha d(p, y)$. Then $d(m_1, m_2) \leq \alpha d(x, y)$.*

(iii) *Let $x, y \in X, x \neq y$ and $z, w \in [x, y]$ such that $d(x, z) = d(x, w)$. Then $z = w$.*

(iv) *Let $x, y \in X$. For each $t \in [0, 1]$ there exists unique point $z \in [x, y]$ such that $d(x, z) = t d(x, y)$ and $d(y, z) = (1 - t) d(x, y)$.*

For convenience, from now onwards we will use the notation $(1-t) \oplus ty$ for the unique point z satisfying (iv).

Lemma 1.2 ([7]). *Let (X, d) be a CAT(0) space. Then*

$$d((1-t)x \oplus ty, z) \leq (1-t)d(x, z) + td(y, z)$$

for all $x, y, z \in X$ and $t \in [0, 1]$.

Lemma 1.3 ([15]). *Let p, x, y be points of a CAT(0) space X, let $\alpha \in [0, 1]$. Then*

$$d((1-\alpha)p \oplus \alpha x, (1-\alpha)p \oplus \alpha y) \leq \alpha d(x, y).$$

The following Lemma is a generalization of (CN) inequality.

Lemma 1.4 ([7]). *Let (X, d) be a CAT(0) space. Then*

$$d((1-t)x \oplus ty, z)^2 \leq (1-t)d(x, z)^2 + td(y, z)^2 - t(1-t)d(x, y)^2$$

for all $x, y, z \in X$ and $t \in [0, 1]$.

Let $\{x_n\}$ be a bounded sequence in a CAT(0) space X. For each $x \in X$, we set
$$r(x, \{x_n\}) = \limsup_{n \to \infty} d(x, x_n).$$
The **asymptotic radius** $r(\{x_n\})$ of $\{x_n\}$ is given by
$$r(\{x_n\}) = \inf\{r(x, \{x_n\}) : x \in X\}.$$
And the **asymptotic center** $A(\{x_n\})$ of $\{x_n\}$ is the set
$$A(\{x_n\}) = \{x \in X : r(\{x_n\}) = r(x, \{x_n\})\}.$$
Therefore, the following equivalence holds for any point $u \in X$:

(1.1) $\quad u \in A(\{x_n\}) \Leftrightarrow \limsup_{n \to \infty} d(u, x_n) \leq \limsup_{n \to \infty} d(x, x_n)$

for all $x \in X$. It is known (see, e.g., [8], Proposition 7) that in a CAT(0) space, $A(\{x_n\})$ consists of exactly one point.

We now give the definition of Δ-convergence in a CAT(0) space.

Definition 1.1 ([13]). *A sequence $\{x_n\}$ in X is said to be Δ-convergent to $x \in X$ if x is the unique asymptotic center of $\{u_n\}$ for every subsequence $\{u_n\}$ of $\{x_n\}$. In this case we write*

(1.2) $\quad \Delta - \lim_n x_n = x$ *and call x the $\Delta - $ limit of $\{x_n\}$.*

We denote, $\omega_\Delta(x_n) = \cup\{A(\{u_n\})\}$,where the union is taken over all subsequence $\{u_n\}$ of $\{x_n\}$.

Definition 1.2. *Let C be nonempty subset of a CAT(0) space X and $T : C \to X$ be a mapping. Then T is called **nonexpansive** if for each $x, y \in C$,*

(1.3) $\quad d(Tx, Ty) \leq d(x, y).$

A point $x \in C$ is called a fixed point of T if $x = Tx$. We denote with $F(T)$ the set of fixed points of T.

Lemma 1.5 ([7]). *Let (X, d) be a CAT(0) space. Then*

(i) *Every bounded sequence in X has a $\Delta - $ convergent subsequence.*

(ii) *If C is a closed convex subset of X and if $\{x_n\}$ is a bounded sequence in C, then the asymptotic center of $\{x_n\}$ is in C.*

(iii) *If C is a closed convex subset of X and if $T : C \to X$ is nonexpansive mapping, then the conditions , $\{x_n\}$ is $\Delta - $ convergent to x and $d(x_n, T(x_n)) \to 0$, imply $x \in C$ and $T(x) = x$.*

Let C be a nonempty subset of a Banach space X and $T, S : C \to C$ be two mappings. The Picard iterative procedure is defined by the sequence $\{x_n\}$:

(1.4) $$\begin{cases} x_1 = x \in C \\ x_{n+1} = Tx_n, n \in N \end{cases}$$

In 1953, Mann[18] defined the following iterative procedure:

(1.5) $$\begin{cases} x_1 = x \in C \\ x_{n+1} = (1 - a_n)x_n + a_n Tx_n, n \in N \end{cases}$$

where $\{a_n\}$ is in $(0, 1)$. It is known that Picard iteration scheme converges for contractions but may not converge for nonexpansive mappings whereas Mann iterative procedure converges for nonexpansive mappings as well.

The sequence $\{x_n\}$ defined by

(1.6) $$\begin{cases} x_1 = x \in C \\ x_{n+1} = (1 - a_n)x_n + a_n Ty_n \\ y_n = (1 - b_n)x_n + b_n Tx_n, n \in N \end{cases}$$

where $\{a_n\}$ and $\{b_n\}$ are in $(0, 1)$, is known as the Ishikawa[10] iterative procedure.

In 2007, Agarwal et. al.[1] introduced the following iterative procedure:

(1.7) $$\begin{cases} x_1 = x \in C \\ x_{n+1} = (1 - a_n)Tx_n + a_n Ty_n \\ y_n = (1 - b_n)x_n + b_n Tx_n, n \in N \end{cases}$$

where $\{a_n\}$ and $\{b_n\}$ are in $(0, 1)$. The iterative procedure (1.7) is independent of (1.6) (and hence of (1.5)). They showed that this procedure converges at a rate same as that of Picard iteration and faster than Mann (1.5) for contractions. It is easy to see on the similar lines that iterative procedure (1.7) also converges faster than the Ishikawa iterative procedure (1.6)

The above procedures deal with one mapping only. The case of two mappings in iterative procedures was firstly studied by Das and

Debata[6] on the pattern of the Ishikawa iterative procedure:

(1.8) $$\begin{cases} x_1 = x \in C \\ x_{n+1} = (1-a_n)x_n + a_n T y_n \\ y_n = (1-b_n)x_n + b_n S x_n, n \in N \end{cases}$$

where $\{a_n\}$ and $\{b_n\}$ are in $(0,1)$. This iterative procedure reduces to the Ishikawa iterative procedure (1.6) when $S = T$ and to Mann iterative procedure (1.5) when $S = I$.

Yao and Chen[22] studied the following iterative procedure:

(1.9) $$\begin{cases} x_1 = x \in C \\ x_{n+1} = a_n x_n + b_n T x_n + c_n S x_n, n \in N \end{cases}$$

where $\{a_n\}, \{b_n\}$ and $\{c_n\}$ are in $(0,1)$ and $a_n + b_n + c_n = 1$. We note that (1.9) reduces to Mann iterative procedure (1.5) when $T = I$ or $S = I$.

In 2010, Khan et. al.[12] modified the iterative procedure (1.7) to case of two mappings as follows:

(1.10) $$\begin{cases} x_1 = x \in C \\ x_{n+1} = (1-a_n)Tx_n + a_n S y_n \\ y_n = (1-b_n)x_n + b_n T x_n, n \in N \end{cases}$$

where $\{a_n\}$ and $\{b_n\}$ are in $(0,1)$. This iterative procedure reduces to the Agarwal et. al iterative procedure (1.7) when $S = T$ and to Mann iterative procedure (1.5) when $T = I$. They showed that this procedure converges at a rate same as that of Picard (1.4) and Agarwal et. al (1.7) but faster than (1.5), (1.8) and (1.9) for contractions.

The iterative procedure (1.10) is independent of both (1.8) and (1.9). Also neither of (1.8) and (1.9) reduces to (1.7) nor conversely. It means that results proved by (1.8) and (1.9) do not include results proved by (1.7).

Clearly (1.7) does not reduces to (1.5) but (1.10) does. This implies that (1.10) not only covers the results proved by (1.7) but also by (1.5) which are not covered by (1.7).

In 2011, Khan and Abbas[11] modified (1.7) for a nonexpansive mapping $T : C \to C$ (where C be a nonempty subset of a CAT(0) space X) as follows:

(1.11) $$\begin{cases} x_1 = x \in C \\ x_{n+1} = (1-a_n)Tx_n \oplus a_n T y_n \\ y_n = (1-b_n)x_n \oplus b_n T x_n, n \in N \end{cases}$$

where $\{a_n\}$ and $\{b_n\}$ are in $(0,1)$.

Let C be a nonempty subset of a CAT(0) space X and $T, S : C \to C$ be two mappings. We now modify (1.10) in CAT(0) spaces as follows:

(1.12) $$\begin{cases} x_1 = x \in C \\ x_{n+1} = (1-a_n)Tx_n \oplus a_n S y_n \\ y_n = (1-b_n)x_n \oplus b_n T x_n, n \in N \end{cases}$$

where $\{a_n\}$ and $\{b_n\}$ are in $(0,1)$.

The aim of this paper is to study Khan et. al. iterative procedure (1.12) for approximating common fixed points of two nonexpansive mappings in the setting of CAT(0) spaces. This iterative procedure extends Agarwal et. al. iterative procedure (1.11) studied by Khan and Abbas[11] to the case of two mappings. Also it is faster than both Ishikawa type iterative procedure (1.8) and the one studied by Yao and Chen (1.9). We also obtain some strong and Δ-convergence results for approximating common fixed points of two nonexpansive self mappings using iterative procedure (1.12) in CAT(0) spaces. In the light of Remark.1.1 it is also noted that our results in CAT(0) spaces can be applied to CAT(κ) spaces with $\kappa \leq 0$.

2. Main Results

In this section we obtain strong and Δ-convergence theorems of iterative procedure (1.12). In the sequel F denotes the set of common fixed points of the mappings T and S.

Lemma 2.1. *Let C be a nonempty closed convex subset of a CAT(0) space X and $T, S : C \to C$ be two nonexpansive mappings. Let $\{x_n\}$ be defined by iterative procedure (1.12). Let $\{a_n\}$ and $\{b_n\}$ be such that $0 < a \leq a_n, b_n \leq b < 1$ for all $n \in N$ and for some a, b. Then*

(i) $\lim_{n \to \infty} d(x_n, q)$ *exists for all* $q \in F$.

(ii) $\lim_{n \to \infty} d(x_n, Tx_n) = 0 = \lim_{n \to \infty} d(x_n, Sx_n)$.

Proof. Let $q \in F$. Then by Lemma 1.2,

$$(2.1) \quad \begin{aligned} d(x_{n+1}, q) &= d((1-a_n)Tx_n \oplus a_n Sy_n, q) \\ &\leq (1-a_n)d(Tx_n, q) + a_n d(Sy_n, q) \\ &\leq (1-a_n)d(x_n, q) + a_n d(y_n, q). \end{aligned}$$

But

$$(2.2) \quad \begin{aligned} d(y_n, q) &= d((1-b_n)x_n \oplus b_n Tx_n, q) \\ &\leq (1-b_n)d(x_n, q) + b_n d(Tx_n, q) \\ &\leq (1-b_n)d(x_n, q) + b_n d(x_n, q) \\ &= d(x_n, q). \end{aligned}$$

Combining (2.1) and (2.2), we have

$$(2.3) \quad d(x_{n+1}, q) \leq d(x_n, q).$$

Thus, $\{d(x_n, q)\}$ is decreasing and hence $\lim_{n \to \infty} d(x_n, q)$ exists for all $q \in F$. This proves part (i). Let

$$(2.4) \quad \lim_{n \to \infty} d(x_n, q) = c.$$

By (2.1), we have $d(x_{n+1}, q) \leq (1-a_n)d(x_n, q) + a_n d(y_n, q)$. Thus,

$$a_n d(x_n, q) \leq d(x_n, q) + a_n d(y_n, q) - d(x_{n+1}, q),$$

that is,

$$\begin{aligned} d(x_n, q) &\leq d(y_n, q) + \frac{1}{a_n}[d(x_n, q) - d(x_{n+1}, q)] \\ &\leq d(y_n, q) + \frac{1}{a}[d(x_n, q) - d(x_{n+1}, q)]. \end{aligned}$$

This gives

$$\liminf_{n \to \infty} d(x_n, q) \leq \liminf_{n \to \infty} d(y_n, q) + \liminf_{n \to \infty} \frac{1}{a}[d(x_n, q) - d(x_{n+1}, q)]$$

so that

$$(2.5) \quad c \leq \lim_{n \to \infty} d(y_n, q).$$

By (2.2) and (2.4), we get $\limsup_{n \to \infty} d(y_n, q) \leq c$. Combining it with (2.5), we have

$$(2.6) \quad \lim_{n \to \infty} d(y_n, q) = c.$$

Now, by Lemma 1.4,
$$\begin{aligned}d(y_n,q)^2 &= d((1-b_n)x_n \oplus b_n Tx_n, q)^2 \\ &\leq (1-b_n)d(x_n,q)^2 + b_n d(Tx_n,q)^2 - b_n(1-b_n)d(x_n,Tx_n)^2 \\ &\leq (1-b_n)d(x_n,q)^2 + b_n d(x_n,q)^2 - b_n(1-b_n)d(x_n,Tx_n)^2 \\ &\leq d(x_n,q)^2 - b_n(1-b_n)d(x_n,Tx_n)^2\end{aligned}$$

Using (2.4) and (2.6), we get $\limsup_{n\to\infty} d(x_n, Tx_n) \leq 0$. Thus,

(2.7) $$\lim_{n\to\infty} d(x_n, Tx_n) = 0.$$

Again using Lemma 1.4, we get
$$\begin{aligned}d(x_{n+1},q)^2 &= d((1-a_n)Tx_n \oplus a_n Sy_n, q)^2 \\ &\leq (1-a_n)d(Tx_n,q)^2 + a_n d(Sy_n,q)^2 - a_n(1-a_n)d(Tx_n,Sy_n)^2 \\ &\leq (1-a_n)d(Tx_n,q)^2 + a_n d(y_n,q)^2 - a_n(1-a_n)d(Tx_n,Sy_n)^2.\end{aligned}$$

Using (2.4) and (2.6), we have

(2.8) $$\lim_{n\to\infty} d(Tx_n, Sy_n) = 0.$$

Now, using Lemma 1.2,
$$\begin{aligned}d(y_n, x_n) &= d((1-b_n)x_n \oplus b_n Tx_n, x_n) \\ &\leq (1-b_n)d(x_n,x_n) + b_n d(Tx_n, x_n).\end{aligned}$$

Using (2.7), we have

(2.9) $$\lim_{n\to\infty} d(y_n, x_n) = 0.$$

Finally,
$$\begin{aligned}d(x_n, Sx_n) &\leq d(x_n, Tx_n) + d(Tx_n, Sy_n) + d(Sy_n, Sx_n) \\ &\leq d(x_n, Tx_n) + d(Tx_n, Sy_n) + d(y_n, x_n).\end{aligned}$$

Using (2.7), (2.8) and (2.9),
$$\lim_{n\to\infty} d(x_n, Sx_n) = 0.$$

$\square$

Theorem 2.2. *Let X, C, T, S, F, $\{a_n\}$, $\{b_n\}$ and $\{x_n\}$ be as in Lemma 2.1. Then, $\{x_n\}$ Δ-converges to a point of F.*

Proof. Let $q \in F$. Then by Lemma 2.1, $\lim_{n\to\infty} d(x_n, q)$ exists for all $q \in F$. Thus $\{x_n\}$ is bounded. As proved in Lemma 2.1, we have $\lim_{n\to\infty} d(x_n, Tx_n) = 0 = \lim_{n\to\infty} d(x_n, Sx_n)$.

Firstly, we show that $\omega_\Delta(x_n) \subset F$. Let $u \in \omega_\Delta(x_n)$, then there exists a subsequence $\{u_n\}$ of $\{x_n\}$ such that $A(\{u_n\}) = \{u\}$. By Lemma 1.5(i), there exists a subsequence $\{v_n\}$ of $\{u_n\}$ such that $\Delta - \lim_n v_n = v$ for some $v \in C$. Then by repeated application of Lemma 1.5(iii) on T and S, we obtain $v \in F$. By above Lemma, $\lim_{n\to\infty} d(x_n, v)$ exists. Now, we claim that $u = v$. Assume on the contrary that $u \neq v$. Then by the uniqueness of asymptotic centers, we have

$$\limsup_{n\to\infty} d(v_n, v) < \limsup_{n\to\infty} d(v_n, u) \leq \limsup_{n\to\infty} d(u_n, u) < \limsup_{n\to\infty} d(u_n, v)$$
$$= \limsup_{n\to\infty} d(x_n, v) = \limsup_{n\to\infty} d(v_n, v),$$

a contradiction. Thus $u = v \in F$ and hence $\omega_\Delta(x_n) \subset F$.

To show that $\{x_n\}$ Δ-converges to a point of F, we show that $\omega_\Delta(x_n)$ consists of exactly one point. Let $\{u_n\}$ be a subsequence of $\{x_n\}$. By Lemma 1.5(i), there exists a subsequence $\{v_n\}$ of $\{u_n\}$ such that $\Delta - \lim_n v_n = v$ for some $v \in C$. Let $A(\{u_n\}) = \{u\}$ and $A(\{x_n\}) = \{x\}$. We have already obtained that $u = v \in F$. Finally, we claim that $x = v$. If not, then existence of $\lim_{n\to\infty} d(x_n, v)$ and uniqueness of asymptotic center imply that

$$\limsup_{n\to\infty} d(v_n, v) < \limsup_{n\to\infty} d(v_n, x) \leq \limsup_{n\to\infty} d(x_n, x) < \limsup_{n\to\infty} d(x_n, v)$$
$$= \limsup_{n\to\infty} d(v_n, v),$$

a contradiction. Thus $x = v \in F$ and hence $\omega_\Delta(x_n) = \{x\}$ Thus, $\{x_n\}$ Δ-converges to a point of F. $\square$

Theorem 2.3. *Let X be a complete CAT(0) space and, C, T, S, F, $\{a_n\}$, $\{b_n\}$, $\{x_n\}$ be as in Lemma 2.1. If $F \neq \phi$ then, $\{x_n\}$ converges strongly to a point of F if and only if $\liminf_{n\to\infty} d(x_n, F) = 0$, where $d(x, F) = \inf\{d(x, p) : p \in F\}$.*

Proof. Necessity is obvious. Conversely, suppose that $\liminf_{n\to\infty} d(x_n, F) = 0$. As proved in Lemma 2.1, we have $d(x_{n+1}, p) \leq d(x_n, p)$ for all $p \in F$. This implies that $d(x_{n+1}, F) \leq d(x_n, F)$ so that $\lim_{n\to\infty} d(x_n, F)$ exists. But by hypothesis $\liminf_{n\to\infty} d(x_n, F) = 0$. Therefore $\lim_{n\to\infty} d(x_n, F) = 0$.

Next, we show that $\{x_n\}$ is a cauchy sequence in C. Let $\varepsilon > 0$ be arbitrarily chosen. Since $\lim_{n \to \infty} d(x_n, F) = 0$, there exists a positive integer n_0 such that

$$d(x_n, F) < \frac{\varepsilon}{4}, \text{ for all } n \geq n_0.$$

In particular, $\inf\{d(x_{n_0}, p) : p \in F\} < \frac{\varepsilon}{4}$. Thus there must exist $p' \in F$ such that $d(x_{n_0}, p') < \frac{\varepsilon}{2}$. Now, for all $m, n \geq n_0$ we have

$$\begin{aligned} d(x_{n+m}, x_n) &\leq d(x_{n+m}, p') + d(x_n, p') \\ &\leq 2d(x_{n_0}, p') \\ &< 2\frac{\varepsilon}{2} = \varepsilon. \end{aligned}$$

Hence $\{x_n\}$ is a Cauchy sequence in a closed subset C of a complete CAT(0) space X and so it must converge to a point q in C. Now, $\lim_{n \to \infty} d(x_n, F) = 0$, gives that $d(q, F) = 0$. Since F is closed, so we have $q \in F$. □

Fukhar-ud-din and Khan[9] introduced the condition($\acute{A}$) as follows: Two mappings $T, S : C \to C$ are said to satisfy the condition($\acute{A}$) if there exists a nondecreasing function $f : [0, \infty) \to [0, \infty)$ with $f(0) = 0, f(r) > 0$ for all $r \in (0, \infty)$, such that either $f(d(x, F)) \leq d(x, Tx)$ or $f(d(x, F)) \leq d(x, Sx)$ for all $x \in C$, where $d(x, F) = \{d(x, p) : p \in F\}$

If we take $S = T$ in this condition, then it reduces to condition(A) of Senter and Doston[20].

Applying Theorem 2.3, we obtain strong convergence of the iterative procedure (1.12) under the condition($\acute{A}$) as follows:

Theorem 2.4. *Let X be a complete CAT(0) space and C, T, S, F, $\{a_n\}$, $\{b_n\}$, $\{x_n\}$ be as in Lemma 2.1. Suppose that T, S satisfies the condition($\acute{A}$) and $F \neq \phi$. Then, sequence $\{x_n\}$ converges strongly to a point of F.*

Proof. We have proved in Lemma 2.1 that $\lim_{n \to \infty} d(x_n, p)$ exists for all $p \in F$. Let this limit be c. As proved in Lemma 2.1, we have $d(x_{n+1}, p) \leq d(x_n, p)$ for all $p \in F$. This gives that $\inf_{p \in F} d(x_{n+1}, p) \leq \inf_{p \in F} d(x_n, p)$, which means that $d(x_{n+1}, F) \leq d(x_n, F)$, so that $\lim_{n \to \infty} d(x_n, F)$ exists. Again using Lemma 2.1, we have $\lim_{n \to \infty} d(x_n, Tx_n) = 0 = \lim_{n \to \infty} d(x_n, Sx_n)$.

From the condition($\acute{A}$), either
$$\lim_{n\to\infty} f(d(x_n, F)) \leq d(x_n, Tx_n) = 0, \text{ or } \lim_{n\to\infty} f(d(x_n, F)) \leq d(x_n, Sx_n) = 0.$$
Hence, $\lim_{n\to\infty} f(d(x_n, F)) = 0$. Since $f : [0, \infty) \to [0, \infty)$ is a nondecreasing function satisfying $f(0) = 0, f(r) > 0$ for all $r \in (0, \infty)$, therefore we have $\lim_{n\to\infty} d(x_n, F) = 0$. Now, all the conditions of Theorem 2.3 are satisfied, therefore by its conclusion $\{x_n\}$ converges strongly to a point of F. □

Remark 2.1. *If we take $S = T$ in Theorems 2.2, 2.3 and 2.4 we obtain theorems 1, 2 and 3 of Khan and Abbas[11].*

References

[1] Agarwal, R. P., Regan, D. O. and Sahu, D. R., *Iterative construction of fixed points of nearly asymptotically nonexpansive mappings*, J. Nonlinear Convex Anal., **8**, (1), (2007), 61–79.

[2] Beg, I. and Abbas, M., *An iterative process for a family of asymptotically quasi-nonexpansive mappings in CAT(0) spaces*, Novi Sad J. Math., **41**,(2) ,(2011), 149–157.

[3] Bridson, M. and Haefiger, A., *Metric Spaces of Non-Positive Curvature*, Springer- Verlag, Berlin, Heidelberg,(1999).

[4] Bruhat, F. and Tits, J., *Groupes rductifs sur un corps local. I. Donnes radicielles values*, Inst. Hautes tudes Sci. Publ. Math., **41**, (1972), 5–251.

[5] Chaoha, P. and Phon-on, A., *A note on fixed point set in CAT(0) spaces*, J. Math. Anal. Appl., **320**, (2), (2006), 983–987.

[6] Das, G. and Debata, J. P., *Fixed points of quasi-nonexpansive mappings*, Indian J. Pure. Appl. Math., **17**, (1986), 1263–1269.

[7] Dhompongsa, S. and Panyanak, B., *On Δ-convergence theorems in CAT(0) spaces*, Comput. Math. Appl., **56**, (2008), 2572–2579.

[8] Dhompongsa, S., Kirk, W. A. and Sims, B., *Fixed points of uniformly lipschitzian mappings*, Nonlinear Anal. : TMA, **65**, (2006), 762–772.

[9] Fukhar-ud-din, H. and Khan, S. H., *Convergence of iterates with errors of asymptotically quasi nonexpansive mappings and applications*, J. Math. Anal. Appl., **328**, (2007), 821–829.

[10] Ishikawa, S., *Fixed points by a new iteration method*, Proc Amer. Math Soc., **44**, (1974), 147–150.

[11] Khan, S. H. and Abbas, M., *Strong and Δ-convergence of some iterative schemes in CAT(0) spaces*, Computers and Mathematics with Applications, **61**, (2011), 109–116.

[12] Khan, S. H. and Kim, J. K., *Common fixed points of two nonexpansive mappings by a modified faster iteration scheme*, Bull. Korean Math. Soc., **47**,(5), (2010), 973–985.

[13] Kirk, W. A. and Panyanak, B., *A concept of convergence in geodesic spaces*, Nonlinear Anal., **68**, (2008), 3689–3696.

[14] Kirk, W. A., *Geodesic geometry and fixed point theory II*, In International Conference on Fixed Point Theory and Applications, Yokohama Publ., Yokohama, (2004), 113–142.

[15] Kirk, W. A., *Geodesic geometry and fixed point theory*, , In Seminar of Mathematical Analysis (Malaga/Seville, 2002/2003), Colecc. Abierta, 64, Univ. Sevilla Secr. Publ., Seville,(2003), 195–225.

[16] Laokul, T. and Panyanak, B., *Approximating fixed points of nonexpansive mappings in CAT(0) spaces*, Int. Journal of Math. Anal., **3**,(27), (2009), 1305–1315.

[17] Lim, T.C., *Remarks on some fixed point theorems*, Proc. Amer. Math. Soc., **60** ,(1976), 179–182.

[18] Mann, W. R., *Mean value methods in iteration*, Proc. Amer. Math. Soc., **4** ,(1953), 506–510.

[19] Panyanak, B. and Laokul, T., *On the Ishikawa itertaion process in CAT(0) spaces*, Bull. of the Iranian Math. Soc., **37**,(4), (2011), 205–217.

[20] Senter, H. F. and Dotson, W. G., *Approximating fixed points of nonexpansive mappings*, Proc. Amer. Math. Soc., **44**, (1974), 375–380.

[21] Shahzad, N., *Fixed point results for multimaps in CAT(0) spaces*, Top. and Appl., **156**,(5), (2009), 997–1001.

[22] Yao, Y. and Chen, R., *Weak and strong convergence of a modified Mann iteration for asymptotically nonexpansive mappings*, Nonlinear Funct. Anal. Appl., **12**,(2), (2007), 307–315.

[1]Department of Mathematics
Vaish College, Rohtak-124001(India)
E-mail: madhumdur@gmail.com
[2]Department of Mathematics
M. D. University, Rohtak-124001(India)
E-mail: chugh.r1@gmail.com